NISTIR 7887

Correction Factors for the NIST Free-Air Ionization Chambers Used to Realize Air Kerma from W-Anode X-Ray Beams

Stephen M. Seltzer
Radiation and Biomolecular Physics Division
Physical Measurement Laboratory

http://dx.doi.org/10.6028/NIST.IR.7887

October 2012

U.S. Department of Commerce
Rebecca M. Blank, Acting Secretary

National Institute of Standards and Technology
Patrick D. Gallagher, Under Secretary of Commerce for Standards and Technology and Director

Correction Factors for the NIST Free-Air Ionization Chambers Used to Realize Air Kerma from W-Anode X-Ray Beams

October 2, 2012

Stephen M. Seltzer

Guest Researcher
Dosimetry Group
Radiation and Biomolecular Physics Division
Physical Measurement Laboratory
National Institute of Standards and Technology
Gaithersburg, MD USA

Abstract

This report details analyses in the development of certain correction factors for free-air ionization chambers used by the National Institute of Standards and Technology to realize air kerma for its measurement standards and calibrations of W-anode x-ray beams. The correction factors dealt with in this report are the electron-loss correction, k_{el}, the photon-scatter correction, k_{sc}, the fluorescence-reabsorption correction, k_{fl}, and the bremsstrahlung-reabsorption correction, k_{br}.

Introduction

This report documents the development of certain correction factors for the parallel-plate free-air chambers (FACs) used by the National Institute of Standards and Technology (NIST) to realize air kerma for W-anode x-ray beams. These FACs are the Lamperti chamber for x-ray beams of 10 kV to 50 kV, the Ritz chamber for x-ray beams of 20 kV to 100 kV, and the Wyckoff-Attix chamber for x-ray beams of 50 kV to 300 kV. Because this report restricts itself to new evaluations of particular FAC-geometry-dependent correction factors, the reader is referred to previous publications (Lamperti *et al.*, 1988; Lamperti and O'Brien, 2001) for more general descriptions of the FACs and the evaluation of the complete list of correction factors, including the historical methods some of which the present report replaces.

The quantity kerma is defined (ICRU, 2011) for ionizing uncharged particles as the quotient of dE_{tr} by dm, where dE_{tr} is the mean sum of the initial kinetic energies of all the charged particles liberated in a mass dm of a material by the uncharged particles incident on dm, thus

$$K = \frac{dE_{tr}}{dm}. \tag{1}$$

In the International System of Units (SI) (BIPM, 2006) kerma has units of J kg^{-1}; the special name for this unit is gray (Gy).

Technically, air kerma, K_{air}, is realized through a measurement of the related quantity exposure. Exposure, X, is the quotient of dq by dm, where dq is the absolute value of the mean total charge of the ions of one sign produced when all the electrons and positrons liberated or created by photons incident on a mass dm of dry air are completely stopped in dry air, thus

$$X = \frac{dq}{dm}. \tag{2}$$

The SI unit of exposure is C kg^{-1} (however, the older unit of Roentgen (R) is still used by some, where 1R $= 2.58\text{x}10^{-4}$ C kg^{-1}). The ionization produced by electrons emitted in atomic/molecular relaxation processes is included in dq. The ionization due to photons emitted by radiative processes (*i.e.*, bremsstrahlung and fluorescence photons) is not to be included in dq. Except for this difference, significant at high energies, the exposure, as defined above, is the ionization analogue of the dry-air kerma.

The quantities exposure and air kerma can be related through use of the mean energy expended in a gas per ion pair formed, divided by the elementary charge, W/e, where W is the mean energy expended in air per ion pair formed when the initial kinetic energy of a charged particle is completely dissipated in the air, and e is the elementary charge. Then

$$K_{air} \approx X \cdot (W/e)/(1 - \overline{g}).$$ (3)

The quantity g is the fraction of the kinetic energy of electrons (and positrons) liberated by the photons that is lost in radiative processes (mainly bremsstrahlung) in air. In Eq. (3), $\overline{g}$ is the mean value of g averaged over the distribution of the air kerma with respect to electron energy.

A schematic of a generic parallel-plate FAC is shown in Fig. 1. The FAC is a shielded container open to the atmosphere into which a portion of the x-ray beam enters through a defining aperture, passes between parallel collecting and high-voltage plates, and leaves through an exit aperture. The dimensions of the components are designed to allow complete slowing down of all electrons produced, which are collected by the electric field across the collecting volume.

The results of a FAC measurement for x-ray beams are then analyzed according to the measurement equation

$$K_{air} \approx (W/e) \frac{Q_{net}}{\rho_{air} V_{eff} (1 - \overline{g})} \prod_i k_i,$$ (4)

where Q_{net} is the measured net ion current (corrected for cosmic-ray and system-generated background), ρ_{air} is the density of air, and V_{eff} is the product of the aperture area and the length of the collecting volume. Equation (4) is an elaboration of combining Eqs. (2) and (3) in which the mass has been replaced by the product $\rho_{air} V_{eff}$ and the application of the various necessary corrections is shown. The radiative-loss correction $\overline{g}$ is very small, effectively zero, for these x-ray beams, and k_i are correction factors for air attenuation, photon scatter, electron loss, etc., within the FAC. It is assumed that the correction factors include also those for humidity, ion recombination, and ambient pressure and temperature (to correct to reference conditions), but these are not the subject of this report (the correction for humidity was discussed in Seltzer et al. (2003), in which the value of 0.998 used by the NIST was verified).

Before continuing, it should be noted that the approximate equality in Eqs. (3) and (4) is used here to reflect the fact that exposure includes the charge of electrons or ions liberated by the incident photons whereas W pertains only to the charge produced during the slowing down of these electrons. Thus the initial charge created by the interaction of the photons should be discounted when transferring the exposure measurement for the determination of air kerma. This difference, although relatively small, tends to become more significant as the photon energy decreases. Additionally, W is not constant as perhaps implied in Eqs. (3) and (4), but is known to increase at low energies (ICRU, 1979). At energies for which the variation of W with energy becomes important, one should consider also the effect of this increase. The initial-ion correction has been considered by Büermann *et al.* (2006), and by Takata and Begum (2008). The NIST does not yet include this correction; it appears to be relatively small, *i.e.*, rather close to unity, but does seem to become more significant, approaching about 0.995 for standard x-ray beams with the lowest mean energies.

Correction Factors

The correction factors of concern in this report are mainly those for electron loss, photon scatter, fluorescence reabsorption, and bremsstrahlung reabsorption. The first two of these have been evaluated and used for more than 50 years (see Lamperti *et al.*, 1988; Lamperti and O'Brien, 2001; and references therein), but the advent of reliable radiation-transport Monte Carlo calculations has allowed more accurate and consistent data, including that for the latter two correction factors, which although small had not been considered until about a decade ago.

Electron-loss correction, k_{el}. Energetic electrons can leave (and enter) the collection volume, with only a portion of their energy expended in ionization being collected. The collecting volume is defined by the area of the collecting plate (collector length × collector width in Fig. 1) and the electrode separation.

Photon-scatter correction, k_{sc}. Ionization produced by electrons resulting from photons scattered out of the aperture-defined beam is not included in the definitions of exposure and of air kerma.

Fluorescence-reabsorption correction, k_{fl}. The ionization collected due to reabsorption of fluorescence photons is not included in the definitions of exposure and of air kerma.

Bremsstrahlung-reabsorption correction, k_{br}. The ionization collected due to reabsorption of bremsstrahlung photons is not included in the definitions of exposure and of air kerma. This is associated with but separate from the $(1 - g)$ correction. The NIST currently assumes $\bar{g}$ in Eq. (4) is identically zero for its calibration x-ray beams. If the appropriate small values were to be used, then the use of k_{br} would correct for effects of reabsorption. Although the following analyses will carry along results for k_{br}, the final recommendation will be an estimated uncertainty associated with the assumption that $k_{br}/(1-\bar{g}) = 1$.

Data for k_{el}, k_{sc}, k_{fl}, and k_{br} for Monoenergetic Photons

The source of data for the NIST correction factors is the Monte Carlo calculations reported by Burns (2001) for the FACs of NIST and other national metrology institutes. These calculations were a follow-on to earlier work (Burns, 1999), but using a Version VI of the EGSnrc code that included the transport of fluorescence x rays and of Auger and Coster-Kronig electrons. The critical dimensions used for the NIST FACs are listed in Table 1. In addition to these dimensions, Burns used a gap between electrodes and side walls of 2 cm for the Lamperti and Ritz chambers, and of 5 cm for the Wyckoff-Attix chamber. He assumed electrodes and wall materials to be aluminum and an electrode thickness of 1 cm. The composition of air was taken as that listed in ICRU Report 37 (ICRU, 1984), and the histories of photons and electrons were terminated at a kinetic-energy cut-off of 1 keV. The number of histories was such that Burns states that the relative statistical standard uncertainty of all results is less than 0.01 %.

Burns (1999) performed some sensitivity studies to assess uncertainties arising from his model of the chambers. He estimates the relative standard uncertainties to be 0.05 % for k_{el} and 0.03 % for k_{sc} (k_{fl} and k_{br} were not considered in this earlier study). In a later report, Burns (2003) used the Monte Carlo code PENELOPE to re-evaluate the correction factors for the BIPM FACs. From comparisons among the various results, he estimates a relative uncertainty of 0.03 % for k_{sc} and from 0.03 % to 0.05 % for k_{fl}, depending on the chamber size. Moreover, now with results from different methods of electron transport, he then estimates a relative uncertainty in k_{el} of from 0.01 % to 0.09 %, depending on the x-ray beam energy and thus the chamber used. These latter results are difficult to transfer to the NIST chambers, as the dimensions differ somewhat. However, it seems reasonable to assign to the NIST chambers the relative standard uncertainties due to the computational models employed to be those given in Table 2.

The data supplied by Burns (private communication, 2001) were in the form of tables for each of the NIST FACs, and are reproduced in Table 3. These monoenergetic data were, however, processed by Burns from his Monte Carlo scores in terms of a concatenation of separate classes of energy deposition whose ratios yield the desired quantities. Using his notation and the quantities listed in Table 3, we can recover the essential scores, each as a function of photon energy:

$$P_{dep} = \frac{(1-g)P_{rel}}{k_{eq}},$$

$$P_{col} = \frac{(1-g)P_{rel}}{k_{eq}k_{el}},$$

$$S_{col} = \left(\frac{1}{k_{sc}}-1\right)\frac{(1-g)P_{rel}}{k_{eq}k_{el}}, \tag{5}$$

$$F_{col} = \left(\frac{1}{k_{fl}}-1\right)\frac{(1-g)P_{rel}}{k_{eq}k_{el}k_{sc}},$$

$$G_{col} = \left(\frac{1}{k_{br}}-1\right)\frac{(1-g)P_{rel}}{k_{eq}k_{el}k_{sc}k_{fl}}.$$

The normalization of the quantities given in Eqs. (5) (governed by that of P_{rel}) is unimportant, as we are ultimately interested only in their ratios. Indeed, we see then that the monoenergetic results are simply:

$$k_{el}(E) = \frac{P_{dep}(E)}{P_{col}(E)},$$

$$k_{sc}(E) = \frac{P_{col}(E)}{P_{col}(E) + S_{col}(E)},$$

$$k_{fl}(E) = \frac{P_{col}(E) + S_{col}(E)}{P_{col}(E) + S_{col}(E) + F_{col}(E)}, \qquad (6)$$

$$k_{br}(E) = \frac{P_{col}(E) + S_{col}(E) + F_{col}(E)}{P_{col}(E) + S_{col}(E) + F_{col}(E) + G_{col}(E)}.$$

For a photon spectrum, however, it is formally required to integrate over the numerators and denominators of Eqs. (6). In order to facilitate interpolation for such integration, it was decided to fit and thus smooth the quantities on the right-hand side of Eqs. (6). It was noticed that some of the quantities reconstructed from Eqs. (5) displayed inconsistent results at the lowest energy or energies (possibly due to energy cut-offs in the Monte Carlo calculations); in particular, there should be no contribution by fluorescence re-absorption below the K-edge energy of 3.206 keV for Ar. Therefore in some cases the fitted data were extrapolated down to these lower energies to insure smoothness. Issues of accuracy for these lowest energies are unimportant because x-ray spectra of NIST calibration beams rarely extend down to these energies and, in any case, the relative energy fluence would be quite small.

The original and adopted smoothed data are given in Tables 4 and 5, respectively. Reconstructed results for the monoenergetic-photon correction factors from the smoothed data via Eq. (6) are given in Table 6. These are compared with the original Burns data in Figs. 2 - 4. The modified data are not necessarily better, but they do follow the original rather closely and are indeed smoother. Table 7 gives estimated relative uncertainties due to the smoothing process, calculated as the relative root-mean-square deviation between the data of Tables 3 and 6. These are combined (in quadrature) with the results that were given in Table 2 for the estimated total relative standard uncertainties <u>inherent</u> in the basic data adopted for calculating correction factors for the NIST FACs, and these are listed in Table 8. That is, it is assumed that any additional uncertainty is to be contributed by that in the input spectra for the incident x-ray beam.

As regards k_{br}, one can form the product $k_{br}/(1-g)$ from Burns' Monte Carlo results in Table 3, and investigate its difference from unity. Such results suggest an estimate of the relative standard uncertainties associated with the assumption that $k_{br}/(1-\bar{g}) = 1$ of 0.02 % for the Lamperti chamber, 0.02 % for the Ritz chamber, and 0.03 % for the Wyckoff-Attix chamber. These have been added to Table 8 for completeness.

Results for the NIST Standard Beam Qualities

There has developed a fair amount of information on the W-anode x-ray fluence spectra that the NIST uses in its measurement standards and calibration services (see, Seelentag et al., 1979; Eisenhower et al., 1983; Iles, 1987; Peaple et al., 1989; Laitano et al., 1989; 1991; Ankerhold, 2000), as well as the tools to estimate such fluence spectra for a wide range of conditions (see, e.g., Cranley et al., 1997). C.G. Soares of the NIST collected much of this information in digital form and made it available (private communication, 1994). These spectra, including the NIST traditional and ISO beams (ISO, 1996), number some 200 versions from independent measurements (from the institutions GSF, Harwell, ENEA, PTB, and NIST) and calculations (from the NRPB), thus often providing multiple versions (up to 5) for many of the NIST beam qualities.

The smoothed data in Table 5 were integrated over these available fluence spectra, the ratios of the integrated values were taken as indicated in Eq. (6), and the results from multiple versions of the spectra were averaged. The adopted results are listed in Table 9 for the traditional NIST beams qualities, in Table 10 for the ISO beam qualities, and in Table 11 for the BIPM/CCRI beam qualities used for international comparisons. There were no fluence spectra available for the more recently added NIST beam qualities M80 and M120, so the correction factors were obtained by interpolation as a function of the half-value layer for the Ritz and for the Wyckoff FACs.

The variations in results from multiple spectra for the same beam quality suggest that a reasonable estimate of the relative standard uncertainty of the results is 20 % of the (absolute) difference of the correction factor from unity. Because the correction factors are so close to unity, this estimated relative standard uncertainties of the correction factors themselves due to assumed fluence spectrum is never more than about 0.01 %. This then suggests that the uncertainties listed in Table 8 are appropriate for the final results in Tables 9 and 10.

Final Remarks

This work is the culmination of efforts started in 2001 to produce correction factors for the NIST FACs. The results produced in various analyses over the last decade are rather consistent, but the present results should be considered as the most complete, with a perhaps more careful uncertainty analysis. The analyses have also included the correction for attenuation within the FAC, k_{at}, using standard reference data for the photon attenuation coefficients, presumably consistent with the Monte Carlo calculations by Burns. Because of ambiguity on the role of coherent scattering in the analytical calculations of the attenuation, and on the uncertainty of the underlying photon-interaction cross sections, the calculated k_{at} values were intended only to confirm measured values, which could be readily obtained (although not without some difficulty). Thus these values are not presented here, although agreement with measured values is reasonably good.

Acknowledgements

This work was conducted over the years in close collaboration with Michelle O'Brien, the NIST expert for air-kerma measurement standards for x-ray beams and ultimate user of these results. Of course, this work would not have been possible without the Monte Carlo results and background explanations supplied by David Burns of the BIPM, nor without the collection of x-ray-beam fluence spectra provided by Christopher Soares of the NIST. The efforts of the author were partially supported through contract SB1341-11-RQ-0398 with the Dosimetry Group, Radiation and Biomolecular Physics Division, National Institute of Standards and Technology.

References

Ankerhold, A. (2000). *Catalogue of X-Ray Spectra and their Characteristic Data - ISO and DIN Radiation Qualitites, Therapy and Diagnostic Radiation Qualities, Unfiltered X-Ray Spectra, PTB Report* PTB-Dos-34, Physikalisch-Technische Bundesanstalt, Braunschweig, Germany.

Büermann, I., Grosswendt, B.,. Kramer, H.-M., Selbach, H.-J., Gerlach, M., Hoffmann, M., and Krumrey, M. (2006). "Measurement of the x-ray mass energy-absorption coefficient using 3 keV to 10 keV synchrotrton radiation," *Phys. Med. Biol.* **51**, 5125–5150.

BIPM (2006). Bureau International des Poids et Mesures, *Le Systéme International d'Unités (SI), The Initernational System of Units (SI)*, 8th edition (Bureau International des Poids et Mesures, Sèvres).

Burns, D.T. (1999). "Consistent set of calculated values for electron-loss and photon-scatter corrections for parallel-plate free-air chambers," *CCRI(I)*/99-4, BIPM, Sèvres, France.

Burns, D.T. (2001). "Free-air chamber correction factors for electron loss, photon scatter, fluorescence and bremsstrahlung," *CCRI(I)*/01-36, BIPM, Sèvres, France.

Cranley,K., Gilmore, B.J., Fogarty, G.W.A., and Desponds, L. (1997). *Catalogue of Diagnostic X-ray Spectra and Other Data*, Report No. 78, Institute of Physics and Engineering in Medicine, York, UK.

Eisenhower, E.H., Ehrlich, M., Soares, C., Schima, F.J., and Seltzer, S.M. (1983). *Quality Assurance for Measurements of Ionizing Radiation* (Annual Report for FY 1982), USNRC Report NUREG/CR-3120, Washington, D.C.

ICRU (1979). International Commission on Radiation Units and Measurements, *Average Energy Required to Produce an Ion Pair,* ICRU Report 31 (International Commission on Radiation Units and Measurements, Bethesda, MD).

ICRU (2011). International Commission on Radiation Units and Measurements, *Fundamental Quantities and Units for Ionizing Radiation,* ICRU Report 85a (International Commission on Radiation Units and Measurements, Bethesda, MD).

Iles, W.J. (1987). *The Computation of Bremsstrahlung X-Ray Spectra over an Energy Range 15 keV to 300 keV.* Document NRPB-R204, National Radiological Protection Board, Chilton, UK.

ISO (1996). International Organization for Standardization, *X and gamma reference radiations for calibrating dosimeters and dose rate meters and for determining their responses as a function of photon energy -- Part 1.: Radiation characteristics and production methods*, ISO/IS 4037-1:1996 (International Organization for Standardization, Geneva).

Laitano, R.F. Pani, R. Pellegrini, R., and Toni, M.P. (1989). *Energy Distributions and Air Kerma Rates of ISO and BIPM Reference Filtered X-Radiations*, Technical Report ENEA RT/PAS(89), ENEA (National Agency for Atomic Energy), Rome, Italy.

Laitano, R.F. Pani, R. and Pellegrini, R.(1991). "Determination od x-ray spectra and of the scattered component up to 300 kV," *Med. Phys.* **18**, 934–938.

Lamperti, P.J., Loftus, T.P., and Loevinger, R. (1988). *NBS Measurement Services: Calibration of X-Ray and Gamma-Ray Measuring Instruments*, National Bureau of Standards Special Publication 250-16, Gaithersburg, MD.

Lamperti, P.J., and O'Brien, M. (2001). *NIST Measurement Services: Calibration of X-Ray and Gamma-Ray Measuring Instruments*, National Institute of Standards and Technology Special Publication 250-58, Gaithersburg, MD.

Peaple, L.H.J., Birch, R., and Marshall, M. (1989). *Measurement of the ISO Series of Filtered Radiations,* Report AERE-R-13424, Atomic Energy Research Establishment, Harwell, England.

Seelentag, W.W., Panzer, W., Drexler, G., Platz, L., and Santner, F. (1979). *A Catalogue of Spectra for the Calibration of Dosemeters*, GSF Bericht 560. Gesellschaft fur Strahlen und Umweltforschung MBH, Munchen, Germany.

Seltzer, S.M., Lamperti, P.J., Loevinger, R., Mitch, M.G., Weaver, J.T., and Coursey, B.M. (2003). "New National Air-Kerma-Strength Standards for ^{125}I and ^{103}Pd Brachytherapy Seeds," *NIST J. Res.* **108**, 337–358.

Takata, N., and Begum, A. (2008). "Corrections to air kerma and exposure measured with free air ionization chambers for charge of photoelectrons, Compton electrons and Auger electrons," *Radiat. Prot. Dosim.* **130**, 410–418.

Table 1. Critical dimensions for the NIST parallel-plate free-air chambers. All dimensions are in units of cm.

	Lamperti (10 kV – 50 kV)	Ritz (20 kV – 100 kV)	Wyckoff-Attix (50 kV – 300 kV)
Aperture diameter	0.5	1.0	1.0
Collector length	1.0	7.0	10.1
Collector width	5.0	9.0	27.0
Electrode separation	4.0	9.0	27.0
Attenuation length	3.9	12.7	30.8

Table 2. Estimated relative standard uncertainties due to computational models adopted for the NIST FACs.

Correction Factor	Lamperti (10 kV – 50 kV)	Ritz (20 kV – 100 kV)	Wyckoff-Attix (50 kV – 300 kV)
k_{el}	0.05 %	0.05 %	0.09 %
k_{sc}	0.03 %	0.03 %	0.03 %
k_{fl}	0.05 %	0.05 %	0.03 %

Table 3. Original data from Burns' Monte Carlo calculations for NIST FACs. Note that values of k_{el} shown as 1.0 indicate that Burns evidently imposed this lower limit rather than giving the results from the ratio shown in Eq. (6). The various quantities are indicated in the text. The large number of significant figures is to preserve the data as transmitted.

a. Lamperti chamber.

E/keV	$(1-g)^{-1}$	k_{eq}	k_{el}	k_{sc}	k_{fl}	k_{br}	P_{rel}
2	1.0000000	0.9997988	1.0	0.9980880	1.0000000	1.0000000	4.969E-19
4	1.0000028	0.9999972	1.0	0.9969588	0.9942037	0.9999959	8.651E-18
6	1.0000010	0.9997953	1.0	0.9976122	0.9956900	1.0000000	4.908E-18
8	1.0000120	0.9993245	1.0	0.9980929	0.9964462	0.9999954	3.012E-18
10	1.0000282	1.0005597	1.0	0.9982881	0.9970240	0.9999870	1.702E-18
12	1.0000557	0.9991425	1.0	0.9986930	0.9973521	0.9999813	1.246E-18
14	1.0000841	0.9999159	1.0	0.9985291	0.9975836	0.9999881	9.803E-19
16	1.0000991	0.9986722	1.0	0.9986802	0.9979104	0.9999662	6.502E-19
18	1.0001212	1.0010093	1.0	0.9988290	0.9980633	0.9999629	5.313E-19
20	1.0001339	1.0049185	1.0	0.9989213	0.9982463	0.9999654	3.978E-19
22	1.0001722	0.9986311	1.0	0.9989377	0.9984463	0.9999728	3.338E-19
24	1.0001486	0.9998514	1.0	0.9990345	0.9985611	0.9999669	2.889E-19
26	1.0001149	0.9975016	1.0	0.9989899	0.9986761	0.9999650	2.511E-19
28	1.0001800	0.9975803	1.0	0.9988816	0.9987150	0.9999731	2.227E-19
30	1.0002288	0.9980142	1.0	0.9987269	0.9988919	0.9999514	1.704E-19
32	1.0002138	1.0010957	1.0	0.9990631	0.9988387	0.9999615	1.529E-19
34	1.0002050	1.0002259	1.0002797	0.9990535	0.9990175	0.9999753	1.408E-19
36	1.0002150	1.0043660	1.0016750	0.9989713	0.9990364	0.9999441	1.285E-19
38	1.0003100	1.0024512	1.0056470	0.9990027	0.9991120	0.9999691	1.197E-19
40	1.0001991	1.0002266	1.0135405	0.9987202	0.9991943	0.9999642	1.016E-19
42	1.0002144	0.9989974	1.0267239	0.9991060	0.9992517	0.9999659	9.620E-20
44	1.0003089	1.0018165	1.0445255	0.9990351	0.9992840	0.9999639	9.136E-20
46	1.0002494	0.9993876	1.0702046	0.9989927	0.9993256	0.9999588	8.719E-20
48	1.0002166	0.9935509	1.0982407	0.9990157	0.9992944	0.9999566	8.383E-20
50	1.0002638	0.9995694	1.1251879	0.9988022	0.9993390	0.9999602	7.821E-20

b. Ritz chamber.

E/keV	$(1-g)^{-1}$	k_{eq}	k_{el}	k_{sc}	k_{fl}	k_{br}	P_{rel}
2	1.0000000	1.0000000	1.0	1.0000000	1.0000000	1.0000000	1.156E-21
4	1.0000002	0.9998972	1.0	0.9938854	0.9912199	0.9999996	9.745E-17
6	1.0000000	1.0000000	1.0	0.9947642	0.9934328	1.0000000	1.052E-16
8	1.0000017	0.9997311	1.0	0.9953721	0.9945537	0.9999994	7.485E-17
10	1.0000300	0.9999700	1.0	0.9962687	0.9953324	0.9999838	4.523E-17
12	1.0000530	0.9999470	1.0	0.9966030	0.9959413	0.9999818	3.365E-17
14	1.0000669	0.9999331	1.0	0.9966397	0.9965215	0.9999747	2.653E-17
16	1.0000837	0.9999163	1.0	0.9973156	0.9968536	0.9999656	1.790E-17
18	1.0001129	0.9998871	1.0	0.9972494	0.9970578	0.9999562	1.464E-17
20	1.0001163	1.0007919	1.0	0.9974290	0.9973274	0.9999609	1.102E-17
22	1.0001580	1.0007055	1.0	0.9974093	0.9975435	0.9999519	9.271E-18
24	1.0001782	0.9996965	1.0	0.9973497	0.9977324	0.9999524	7.977E-18
26	1.0001747	0.9996822	1.0	0.9974738	0.9978967	0.9999474	6.984E-18
28	1.0001876	1.0006199	1.0	0.9976103	0.9980685	0.9999508	6.196E-18
30	1.0001765	0.9998235	1.0	0.9974863	0.9981725	0.9999493	4.734E-18
32	1.0002152	1.0002538	1.0	0.9976880	0.9983247	0.9999464	4.266E-18
34	1.0002219	1.0010609	1.0	0.9976233	0.9984074	0.9999460	3.902E-18
36	1.0002177	1.0003426	1.0	0.9976737	0.9985103	0.9999538	3.571E-18
38	1.0002195	1.0000822	1.0	0.9977673	0.9985991	0.9999480	3.316E-18
40	1.0002196	0.9997804	1.0	0.9977829	0.9987117	0.9999522	2.828E-18
42	1.0002471	1.0005024	1.0	0.9977390	0.9988409	0.9999419	2.670E-18
44	1.0002448	1.0009426	1.0	0.9977765	0.9988952	0.9999538	2.529E-18
46	1.0002358	1.0009991	1.0	0.9977629	0.9989872	0.9999522	2.432E-18
48	1.0002475	1.0010392	1.0	0.9979882	0.9990457	0.9999486	2.334E-18
50	1.0002315	1.0002255	1.0000036	0.9978334	0.9991229	0.9999506	2.172E-18

b. Ritz chamber (cont'd).

E/keV	$(1-g)^{-1}$	k_{eq}	k_{el}	k_{sc}	k_{fl}	k_{br}	P_{rel}
52	1.0002478	1.0011245	1.0000683	0.9977577	0.9992061	0.9999582	2.085E-18
54	1.0002291	0.9994704	1.0003007	0.9978193	0.9992720	0.9999563	2.038E-18
56	1.0002138	1.0001679	1.0011186	0.9978974	0.9993310	0.9999646	2.002E-18
58	1.0002026	1.0004479	1.0028882	0.9978886	0.9994012	0.9999637	1.984E-18
60	1.0002126	0.9995248	1.0058850	0.9980641	0.9994235	0.9999619	1.968E-18
62	1.0002041	1.0019827	1.0099604	0.9980151	0.9994735	0.9999633	1.913E-18
64	1.0002198	0.9994224	1.0152254	0.9980040	0.9995188	0.9999619	1.912E-18
66	1.0001914	1.0012683	1.0219123	0.9980829	0.9995702	0.9999653	1.924E-18
68	1.0002134	0.9996194	1.0288651	0.9981840	0.9995912	0.9999676	1.936E-18
70	1.0001747	1.0007563	1.0362296	0.9981151	0.9996009	0.9999580	1.952E-18
72	1.0002137	1.0005539	1.0436668	0.9981675	0.9996438	0.9999603	1.973E-18
74	1.0002108	1.0007109	1.0501383	0.9981279	0.9996798	0.9999586	1.995E-18
76	1.0002157	1.0000510	1.0566075	0.9980965	0.9996931	0.9999611	2.025E-18
78	1.0002004	0.9991657	1.0622996	0.9982181	0.9997184	0.9999643	2.050E-18
80	1.0001732	0.9998784	1.0665093	0.9981366	0.9997411	0.9999615	2.083E-18
82	1.0001922	0.9996665	1.0644509	0.9981675	0.9997723	0.9999629	2.069E-18
84	1.0001958	0.9991964	1.0655565	0.9982435	0.9997869	0.9999638	2.100E-18
86	1.0001971	1.0003183	1.0668154	0.9982933	0.9998000	0.9999599	2.139E-18
88	1.0001926	1.0000736	1.0682423	0.9983000	0.9998231	0.9999626	2.183E-18
90	1.0002114	0.9996883	1.0691059	0.9983129	0.9998289	0.9999608	2.231E-18
92	1.0002297	1.0001402	1.0691534	0.9983458	0.9998507	0.9999616	2.276E-18
94	1.0001832	1.0003445	1.0684333	0.9984353	0.9998598	0.9999562	2.324E-18
96	1.0002108	1.0005231	1.0672652	0.9983771	0.9998686	0.9999589	2.370E-18
98	1.0002106	1.0007573	1.0663093	0.9983969	0.9998776	0.9999623	2.425E-18
100	1.0002088	1.0008069	1.0612865	0.9984623	0.9998840	0.9999633	2.453E-18

c. Wyckoff-Attix chamber.

E/keV	$(1-g)^{-1}$	k_{eq}	k_{el}	k_{sc}	k_{fl}	k_{br}	P_{rel}
10	1.0000330	0.9999670	1.0	0.9907690	0.9940463	0.9999788	5.798E-17
20	1.0001388	0.9998612	1.0	0.9931947	0.9965738	0.9999412	1.547E-17
30	1.0001677	0.9999816	1.0	0.9936727	0.9976644	0.9999263	6.699E-18
40	1.0002682	0.9997319	1.0	0.9942618	0.9983533	0.9999264	4.032E-18
50	1.0002738	0.9997263	1.0	0.9948802	0.9988959	0.9999395	3.080E-18
60	1.0002159	0.9997841	1.0	0.9948538	0.9992142	0.9999491	2.807E-18
70	1.0001906	0.9987293	1.0	0.9953227	0.9995257	0.9999541	2.774E-18
80	1.0002135	1.0001245	1.0	0.9958322	0.9996827	0.9999550	2.959E-18
90	1.0002143	0.9999362	1.0001646	0.9958384	0.9997936	0.9999433	3.174E-18
100	1.0002133	0.9994791	1.0014520	0.9965291	0.9998543	0.9999365	3.501E-18
110	1.0001829	0.9995988	1.0035905	0.9966065	0.9998860	0.9999414	3.869E-18
120	1.0002313	0.9986789	1.0064862	0.9967747	0.9999239	0.9999418	4.291E-18
130	1.0002146	0.9994416	1.0082342	0.9969651	0.9999394	0.9999385	4.728E-18
140	1.0002621	1.0006219	1.0093889	0.9969887	0.9999578	0.9999416	5.208E-18
150	1.0002698	0.9987145	1.0090913	0.9972348	0.9999717	0.9999310	5.624E-18
160	1.0003211	0.9995951	1.0083436	0.9973493	0.9999718	0.9999227	6.104E-18
170	1.0003654	0.9996044	1.0078491	0.9974998	0.9999710	0.9999254	6.574E-18
180	1.0003565	1.0006345	1.0076882	0.9975126	0.9999792	0.9999258	7.083E-18
190	1.0003243	1.0002534	1.0070050	0.9976046	0.9999827	0.9999269	7.570E-18
200	1.0003849	0.9999461	1.0059281	0.9977399	0.9999860	0.9999271	8.036E-18
210	1.0003957	1.0000261	1.0053400	0.9978109	0.9999856	0.9999244	8.550E-18
220	1.0004213	1.0005860	1.0052143	0.9978718	0.9999867	0.9999237	9.049E-18
230	1.0004783	0.9982574	1.0059085	0.9979129	0.9999880	0.9999225	9.535E-18
240	1.0004632	1.0014876	1.0080831	0.9980811	0.9999882	0.9999160	1.005E-17
250	1.0005391	1.0021819	1.0126576	0.9980923	0.9999903	0.9999170	1.054E-17
260	1.0005314	0.9999220	1.0208533	0.9981757	0.9999895	0.9999147	1.102E-17
270	1.0006003	0.9985523	1.0322657	0.9981898	0.9999899	0.9999156	1.150E-17
280	1.0006393	1.0016764	1.0482707	0.9981886	0.9999920	0.9999226	1.202E-17
290	1.0006469	0.9971620	1.0671867	0.9982480	0.9999928	0.9999191	1.248E-17
300	1.0006639	1.0016564	1.0899844	0.9983040	0.9999927	0.9999090	1.299E-17

Table 4. Reconstructed data from Burns' Monte Carlo calculations for NIST FACs, using Eqs. (5). The quantities are defined in the text.

a. Lamperti chamber.

E/keV	P_{dep}	P_{col}	S_{col}	F_{col}	G_{col}
2	4.970E-19	4.970E-19	9.521E-22	0	0
4	8.651E-18	8.651E-18	2.639E-20	5.059E-20	3.561E-23
6	4.909E-18	4.909E-18	1.175E-20	2.130E-20	0
8	3.014E-18	3.014E-18	5.759E-21	1.077E-20	1.386E-23
10	1.701E-18	1.701E-18	2.917E-21	5.086E-21	2.217E-23
12	1.247E-18	1.247E-18	1.632E-21	3.315E-21	2.347E-23
14	9.803E-19	9.803E-19	1.444E-21	2.378E-21	1.172E-23
16	6.510E-19	6.510E-19	8.603E-22	1.365E-21	2.206E-23
18	5.307E-19	5.307E-19	6.222E-22	1.031E-21	1.976E-23
20	3.958E-19	3.958E-19	4.274E-22	6.961E-22	1.373E-23
22	3.342E-19	3.342E-19	3.554E-22	5.206E-22	9.124E-24
24	2.889E-19	2.889E-19	2.792E-22	4.167E-22	9.574E-24
26	2.517E-19	2.517E-19	2.545E-22	3.340E-22	8.836E-24
28	2.232E-19	2.232E-19	2.499E-22	2.875E-22	6.020E-24
30	1.707E-19	1.707E-19	2.176E-22	1.896E-22	8.321E-24
32	1.527E-19	1.527E-19	1.432E-22	1.777E-22	5.891E-24
34	1.407E-19	1.407E-19	1.333E-22	1.385E-22	3.484E-24
36	1.279E-19	1.277E-19	1.315E-22	1.233E-22	7.159E-24
38	1.194E-19	1.187E-19	1.185E-22	1.056E-22	3.676E-24
40	1.016E-19	1.002E-19	1.284E-22	8.090E-23	3.592E-24
42	9.628E-20	9.377E-20	8.391E-23	7.028E-23	3.204E-24
44	9.117E-20	8.728E-20	8.430E-23	6.260E-23	3.160E-24
46	8.722E-20	8.150E-20	8.218E-23	5.506E-23	3.362E-24
48	8.436E-20	7.681E-20	7.568E-23	5.429E-23	3.341E-24
50	7.822E-20	6.952E-20	8.337E-23	4.604E-23	2.770E-24

b. Ritz chamber.

E/keV	P_{dep}	P_{col}	S_{col}	F_{col}	G_{col}
2	1.156E-21	1.156E-21	0	0	0
4	9.746E-17	9.746E-17	5.996E-19	8.686E-19	3.981E-23
6	1.052E-16	1.052E-16	5.537E-19	6.991E-19	0
8	7.487E-17	7.487E-17	3.481E-19	4.119E-19	4.656E-23
10	4.523E-17	4.523E-17	1.694E-19	2.129E-19	7.375E-22
12	3.365E-17	3.365E-17	1.147E-19	1.376E-19	6.186E-22
14	2.653E-17	2.653E-17	8.945E-20	9.292E-20	6.771E-22
16	1.790E-17	1.790E-17	4.818E-20	5.665E-20	6.187E-22
18	1.464E-17	1.464E-17	4.038E-20	4.332E-20	6.447E-22
20	1.101E-17	1.101E-17	2.838E-20	2.958E-20	4.331E-22
22	9.263E-18	9.263E-18	2.406E-20	2.287E-20	4.478E-22
24	7.978E-18	7.978E-18	2.120E-20	1.818E-20	3.815E-22
26	6.985E-18	6.985E-18	1.769E-20	1.476E-20	3.692E-22
28	6.191E-18	6.191E-18	1.483E-20	1.201E-20	3.060E-22
30	4.734E-18	4.734E-18	1.193E-20	8.689E-21	2.411E-22
32	4.264E-18	4.264E-18	9.881E-21	7.172E-21	2.293E-22
34	3.897E-18	3.897E-18	9.284E-21	6.231E-21	2.113E-22
36	3.569E-18	3.569E-18	8.322E-21	5.337E-21	1.656E-22
38	3.315E-18	3.315E-18	7.418E-21	4.661E-21	1.731E-22
40	2.828E-18	2.828E-18	6.284E-21	3.656E-21	1.356E-22
42	2.668E-18	2.668E-18	6.046E-21	3.103E-21	1.555E-22
44	2.526E-18	2.526E-18	5.629E-21	2.800E-21	1.170E-22
46	2.429E-18	2.429E-18	5.446E-21	2.468E-21	1.164E-22
48	2.331E-18	2.331E-18	4.699E-21	2.231E-21	1.201E-22
50	2.171E-18	2.171E-18	4.714E-21	1.910E-21	1.076E-22

b. Ritz chamber (cont'd).

E/keV	P_{dep}	P_{col}	S_{col}	F_{col}	G_{col}
52	2.082E-18	2.082E-18	4.678E-21	1.487E-21	8.718E-23
54	2.039E-18	2.037E-18	4.452E-21	1.340E-21	8.927E-23
56	2.001E-18	1.998E-18	4.210E-21	1.187E-21	7.088E-23
58	1.983E-18	1.976E-18	4.182E-21	1.131E-21	7.186E-23
60	1.969E-18	1.957E-18	3.796E-21	9.978E-22	7.484E-23
62	1.909E-18	1.890E-18	3.760E-21	9.090E-22	6.959E-23
64	1.913E-18	1.884E-18	3.769E-21	8.104E-22	7.193E-23
66	1.921E-18	1.881E-18	3.613E-21	7.712E-22	6.534E-23
68	1.936E-18	1.882E-18	3.424E-21	7.527E-22	6.116E-23
70	1.950E-18	1.882E-18	3.553E-21	6.741E-22	7.914E-23
72	1.971E-18	1.888E-18	3.467E-21	6.086E-22	7.516E-23
74	1.993E-18	1.897E-18	3.557E-21	5.893E-22	7.876E-23
76	2.024E-18	1.916E-18	3.654E-21	5.454E-22	7.462E-23
78	2.051E-18	1.933E-18	3.451E-21	5.077E-22	6.919E-23
80	2.083E-18	1.957E-18	3.653E-21	4.427E-22	7.557E-23
82	2.069E-18	1.940E-18	3.562E-21	4.203E-22	7.206E-23
84	2.101E-18	1.968E-18	3.463E-21	4.011E-22	7.148E-23
86	2.138E-18	2.002E-18	3.422E-21	3.620E-22	8.036E-23
88	2.182E-18	2.043E-18	3.479E-21	3.581E-22	7.653E-23
90	2.231E-18	2.089E-18	3.530E-21	3.185E-22	8.205E-23
92	2.275E-18	2.130E-18	3.529E-21	3.055E-22	8.190E-23
94	2.323E-18	2.175E-18	3.409E-21	2.921E-22	9.552E-23
96	2.368E-18	2.220E-18	3.608E-21	2.788E-22	9.143E-23
98	2.423E-18	2.274E-18	3.652E-21	2.681E-22	8.583E-23
100	2.451E-18	2.307E-18	3.553E-21	1.487E-21	8.474E-23

c. Wyckoff-Attix chamber.

E/keV	P_{dep}	P_{col}	S_{col}	F_{col}	G_{col}
10	5.798E-17	5.798E-17	5.402E-19	3.505E-19	1.248E-21
20	1.547E-17	1.547E-17	1.060E-19	5.355E-20	9.189E-22
30	6.698E-18	6.698E-18	4.265E-20	1.578E-20	4.982E-22
40	4.032E-18	4.032E-18	2.327E-20	6.689E-21	2.988E-22
50	3.080E-18	3.080E-18	1.585E-20	3.422E-21	1.875E-22
60	2.807E-18	2.807E-18	1.452E-20	2.219E-21	1.436E-22
70	2.777E-18	2.777E-18	1.305E-20	1.324E-21	1.282E-22
80	2.958E-18	2.958E-18	1.238E-20	9.428E-22	1.338E-22
90	3.174E-18	3.173E-18	1.326E-20	6.578E-22	1.807E-22
100	3.502E-18	3.497E-18	1.218E-20	5.115E-22	2.229E-22
110	3.870E-18	3.856E-18	1.313E-20	4.410E-22	2.266E-22
120	4.296E-18	4.268E-18	1.381E-20	3.257E-22	2.494E-22
130	4.730E-18	4.691E-18	1.428E-20	2.850E-22	2.896E-22
140	5.203E-18	5.155E-18	1.557E-20	2.180E-22	3.022E-22
150	5.630E-18	5.579E-18	1.547E-20	1.586E-22	3.859E-22
160	6.105E-18	6.054E-18	1.609E-20	1.714E-22	4.691E-22
170	6.574E-18	6.523E-18	1.635E-20	1.894E-22	4.880E-22
180	7.076E-18	7.022E-18	1.751E-20	1.467E-22	5.222E-22
190	7.566E-18	7.513E-18	1.804E-20	1.302E-22	5.508E-22
200	8.033E-18	7.986E-18	1.809E-20	1.118E-22	5.834E-22
210	8.546E-18	8.501E-18	1.865E-20	1.227E-22	6.444E-22
220	9.040E-18	8.993E-18	1.918E-20	1.199E-22	6.877E-22
230	9.547E-18	9.491E-18	1.985E-20	1.140E-22	7.375E-22
240	1.003E-17	9.950E-18	1.913E-20	1.177E-22	8.378E-22
250	1.051E-17	1.038E-17	1.984E-20	1.013E-22	8.633E-22
260	1.102E-17	1.079E-17	1.972E-20	1.139E-22	9.217E-22
270	1.151E-17	1.115E-17	2.022E-20	1.133E-22	9.429E-22
280	1.199E-17	1.144E-17	2.076E-20	9.174E-23	8.876E-22
290	1.251E-17	1.172E-17	2.057E-20	8.445E-23	9.501E-22
300	1.296E-17	1.189E-17	2.020E-20	8.681E-23	1.084E-21

Table 5. The adopted, smoothed data from Burns' Monte Carlo calculations for NIST FACs.

a. Lamperti chamber.

E/keV	P_{dep}	P_{col}	S_{col}	F_{col}	G_{col}
2	1.539E-17	1.539E-17	7.104E-20	1.272E-19	3.256E-23
4	8.684E-18	8.684E-18	2.666E-20	4.921E-20	2.872E-23
6	4.891E-18	4.891E-18	1.152E-20	2.159E-20	2.567E-23
8	2.901E-18	2.901E-18	5.638E-21	1.058E-20	2.313E-23
10	1.842E-18	1.842E-18	3.073E-21	5.706E-21	2.096E-23
12	1.249E-18	1.249E-18	1.840E-21	3.343E-21	1.902E-23
14	8.955E-19	8.955E-19	1.193E-21	2.101E-21	1.726E-23
16	6.718E-19	6.718E-19	8.274E-22	1.402E-21	1.562E-23
18	5.222E-19	5.222E-19	6.071E-22	9.822E-22	1.408E-23
20	4.176E-19	4.176E-19	4.665E-22	7.166E-22	1.262E-23
22	3.419E-19	3.419E-19	3.720E-22	5.400E-22	1.125E-23
24	2.854E-19	2.854E-19	3.055E-22	4.175E-22	9.978E-24
26	2.422E-19	2.422E-19	2.566E-22	3.292E-22	8.801E-24
28	2.084E-19	2.084E-19	2.193E-22	2.636E-22	7.730E-24
30	1.814E-19	1.814E-19	1.898E-22	2.135E-22	6.772E-24
32	1.593E-19	1.593E-19	1.659E-22	1.745E-22	5.930E-24
34	1.411E-19	1.411E-19	1.461E-22	1.437E-22	5.202E-24
36	1.261E-19	1.259E-19	1.295E-22	1.192E-22	4.586E-24
38	1.139E-19	1.132E-19	1.156E-22	9.964E-23	4.078E-24
40	1.042E-19	1.028E-19	1.041E-22	8.413E-23	3.673E-24
42	9.679E-20	9.431E-20	9.470E-23	7.193E-23	3.365E-24
44	9.142E-20	8.747E-20	8.745E-23	6.253E-23	3.152E-24
46	8.742E-20	8.171E-20	8.234E-23	5.555E-23	3.036E-24
48	8.368E-20	7.619E-20	7.954E-23	5.073E-23	3.026E-24
50	7.844E-20	6.972E-20	7.938E-23	4.798E-23	3.140E-24

b. Ritz chamber.

E/keV	P_{dep}	P_{col}	S_{col}	F_{col}	G_{col}
2	2.608E-16	2.608E-16	2.206E-18	2.994E-18	7.171E-22
4	1.663E-16	1.663E-16	1.104E-18	1.411E-18	7.503E-22
6	1.071E-16	1.071E-16	5.738E-19	7.147E-19	7.623E-22
8	7.064E-17	7.064E-17	3.156E-19	3.876E-19	7.547E-22
10	4.807E-17	4.807E-17	1.851E-19	2.240E-19	7.306E-22
12	3.386E-17	3.386E-17	1.159E-19	1.372E-19	6.935E-22
14	2.470E-17	2.470E-17	7.715E-20	8.860E-20	6.474E-22
16	1.862E-17	1.862E-17	5.428E-20	5.993E-20	5.958E-22
18	1.447E-17	1.447E-17	4.005E-20	4.222E-20	5.419E-22
20	1.155E-17	1.155E-17	3.077E-20	3.081E-20	4.882E-22
22	9.439E-18	9.439E-18	2.443E-20	2.316E-20	4.365E-22
24	7.868E-18	7.868E-18	1.992E-20	1.785E-20	3.882E-22
26	6.673E-18	6.673E-18	1.659E-20	1.406E-20	3.439E-22
28	5.742E-18	5.742E-18	1.406E-20	1.126E-20	3.040E-22
30	5.005E-18	5.005E-18	1.209E-20	9.157E-21	2.685E-22
32	4.412E-18	4.412E-18	1.052E-20	7.537E-21	2.374E-22
34	3.930E-18	3.930E-18	9.253E-21	6.268E-21	2.103E-22
36	3.535E-18	3.535E-18	8.222E-21	5.260E-21	1.869E-22
38	3.209E-18	3.209E-18	7.378E-21	4.449E-21	1.668E-22
40	2.941E-18	2.941E-18	6.684E-21	3.790E-21	1.496E-22
42	2.719E-18	2.719E-18	6.113E-21	3.251E-21	1.350E-22
44	2.537E-18	2.537E-18	5.644E-21	2.808E-21	1.226E-22
46	2.388E-18	2.388E-18	5.258E-21	2.441E-21	1.121E-22
48	2.267E-18	2.267E-18	4.942E-21	2.136E-21	1.033E-22
50	2.194E-18	2.193E-18	4.683E-21	1.882E-21	9.589E-23

b. Ritz chamber (cont'd).

E/keV	P_{dep}	P_{col}	S_{col}	F_{col}	G_{col}
52	2.113E-18	2.112E-18	4.470E-21	1.669E-21	8.974E-23
54	2.048E-18	2.047E-18	4.294E-21	1.489E-21	8.467E-23
56	1.998E-18	1.995E-18	4.148E-21	1.338E-21	8.052E-23
58	1.962E-18	1.956E-18	4.025E-21	1.209E-21	7.720E-23
60	1.938E-18	1.927E-18	3.920E-21	1.098E-21	7.458E-23
62	1.925E-18	1.906E-18	3.830E-21	1.003E-21	7.260E-23
64	1.922E-18	1.893E-18	3.752E-21	9.198E-22	7.118E-23
66	1.926E-18	1.886E-18	3.684E-21	8.467E-22	7.026E-23
68	1.938E-18	1.884E-18	3.625E-21	7.817E-22	6.979E-23
70	1.954E-18	1.886E-18	3.575E-21	7.233E-22	6.972E-23
72	1.974E-18	1.891E-18	3.535E-21	6.702E-22	7.001E-23
74	1.995E-18	1.899E-18	3.505E-21	6.215E-22	7.063E-23
76	2.018E-18	1.910E-18	3.487E-21	5.765E-22	7.154E-23
78	2.040E-18	1.923E-18	3.479E-21	5.348E-22	7.269E-23
80	2.063E-18	1.938E-18	3.481E-21	4.961E-22	7.405E-23
82	2.087E-18	1.957E-18	3.492E-21	4.602E-22	7.558E-23
84	2.114E-18	1.980E-18	3.508E-21	4.271E-22	7.723E-23
86	2.144E-18	2.008E-18	3.524E-21	3.969E-22	7.895E-23
88	2.180E-18	2.041E-18	3.536E-21	3.696E-22	8.070E-23
90	2.222E-18	2.080E-18	3.538E-21	3.453E-22	8.241E-23
92	2.270E-18	2.125E-18	3.530E-21	3.240E-22	8.403E-23
94	2.323E-18	2.175E-18	3.515E-21	3.059E-22	8.550E-23
96	2.376E-18	2.227E-18	3.506E-21	2.907E-22	8.675E-23
98	2.421E-18	2.273E-18	3.531E-21	2.784E-22	8.773E-23
100	2.450E-18	2.306E-18	3.641E-21	2.687E-22	8.838E-23

c. Wyckoff-Attix chamber.

E/keV	P_{dep}	P_{col}	S_{col}	F_{col}	G_{col}
3.206	2.035E-16	2.035E-16	2.737E-18	1.710E-18	3.140E-21
5	1.414E-16	1.414E-16	1.691E-18	1.073E-18	2.601E-21
10	5.800E-17	5.800E-17	5.382E-19	3.383E-19	1.606E-21
20	1.548E-17	1.548E-17	1.083E-19	5.682E-20	7.284E-22
30	6.665E-18	6.665E-18	4.110E-20	1.611E-20	4.041E-22
40	4.041E-18	4.041E-18	2.322E-20	6.548E-21	2.656E-22
50	3.105E-18	3.105E-18	1.678E-20	3.381E-21	2.011E-22
60	2.794E-18	2.794E-18	1.408E-20	2.039E-21	1.708E-22
70	2.778E-18	2.778E-18	1.293E-20	1.355E-21	1.592E-22
80	2.928E-18	2.928E-18	1.254E-20	9.573E-22	1.595E-22
90	3.186E-18	3.183E-18	1.256E-20	7.035E-22	1.688E-22
100	3.514E-18	3.506E-18	1.284E-20	5.319E-22	1.859E-22
110	3.889E-18	3.873E-18	1.326E-20	4.120E-22	2.102E-22
120	4.296E-18	4.271E-18	1.378E-20	3.269E-22	2.416E-22
130	4.726E-18	4.691E-18	1.436E-20	2.660E-22	2.796E-22
140	5.172E-18	5.128E-18	1.496E-20	2.223E-22	3.236E-22
150	5.632E-18	5.581E-18	1.556E-20	1.910E-22	3.726E-22
160	6.104E-18	6.049E-18	1.616E-20	1.686E-22	4.249E-22
170	6.585E-18	6.529E-18	1.674E-20	1.527E-22	4.790E-22
180	7.073E-18	7.019E-18	1.731E-20	1.414E-22	5.328E-22
190	7.564E-18	7.514E-18	1.784E-20	1.333E-22	5.847E-22
200	8.056E-18	8.011E-18	1.833E-20	1.274E-22	6.332E-22
210	8.547E-18	8.504E-18	1.876E-20	1.229E-22	6.778E-22
220	9.036E-18	8.991E-18	1.910E-20	1.192E-22	7.182E-22
230	9.527E-18	9.470E-18	1.936E-20	1.158E-22	7.552E-22
240	1.002E-17	9.936E-18	1.955E-20	1.127E-22	7.900E-22
250	1.052E-17	1.038E-17	1.971E-20	1.095E-22	8.241E-22
260	1.102E-17	1.080E-17	1.991E-20	1.061E-22	8.597E-22
270	1.152E-17	1.116E-17	2.019E-20	1.024E-22	8.990E-22
280	1.201E-17	1.146E-17	2.054E-20	9.788E-23	9.443E-22
290	1.248E-17	1.169E-17	2.074E-20	9.169E-23	9.978E-22
300	1.297E-17	1.190E-17	2.017E-20	8.262E-23	1.062E-21

Table 6. The monoenergetic correction factors reconstructed form the adopted, smoothed data from Burns' Monte Carlo calculations for NIST FACs. Values for k_{fl} at energies below 3.206 keV are not given, as they are not expected to have any meaning. The large number of significant figures is for consistency with Table 3.

a. Lamperti chamber.

E/keV	k_{el}	k_{sc}	k_{fl}	k_{br}
2	1.0000000	0.9954059		0.9999979
4	1.0000000	0.9969391	0.9943817	0.9999967
6	1.0000000	0.9976498	0.9956150	0.9999948
8	1.0000000	0.9980605	0.9963734	0.9999921
10	1.0000000	0.9983340	0.9969165	0.9999887
12	1.0000000	0.9985289	0.9973341	0.9999848
14	1.0000000	0.9986697	0.9976620	0.9999808
16	1.0000000	0.9987698	0.9979203	0.9999768
18	1.0000000	0.9988387	0.9981247	0.9999731
20	1.0000000	0.9988842	0.9982890	0.9999699
22	1.0000000	0.9989130	0.9984247	0.9999672
24	1.0000000	0.9989308	0.9985409	0.9999651
26	1.0000000	0.9989417	0.9986442	0.9999638
28	1.0000000	0.9989490	0.9987385	0.9999630
30	1.0000000	0.9989546	0.9988258	0.9999627
32	1.0000000	0.9989599	0.9989075	0.9999629
34	1.0001842	0.9989658	0.9989840	0.9999632
36	1.0018207	0.9989722	0.9990552	0.9999636
38	1.0058194	0.9989797	0.9991215	0.9999640
40	1.0135379	0.9989884	0.9991829	0.9999643
42	1.0262898	0.9989968	0.9992386	0.9999644
44	1.0451052	0.9990013	0.9992863	0.9999640
46	1.0698429	0.9989933	0.9993213	0.9999629
48	1.0983734	0.9989571	0.9993353	0.9999604
50	1.1251489	0.9988627	0.9993131	0.9999550

b. Ritz chamber.

E/keV	k_{el}	k_{sc}	k_{fl}	k_{br}
2	1.0000000	0.9916142		0.9999973
4	1.0000000	0.9934071	0.9916391	0.9999956
6	1.0000000	0.9946713	0.9934059	0.9999930
8	1.0000000	0.9955520	0.9945676	0.9999894
10	1.0000000	0.9961639	0.9953799	0.9999849
12	1.0000000	0.9965896	0.9959779	0.9999797
14	1.0000000	0.9968865	0.9964369	0.9999740
16	1.0000000	0.9970941	0.9968014	0.9999682
18	1.0000000	0.9972401	0.9970991	0.9999628
20	1.0000000	0.9973435	0.9973475	0.9999580
22	1.0000000	0.9974186	0.9975589	0.9999540
24	1.0000000	0.9974751	0.9977420	0.9999509
26	1.0000000	0.9975197	0.9979031	0.9999487
28	1.0000000	0.9975573	0.9980472	0.9999473
30	1.0000000	0.9975907	0.9981782	0.9999466
32	1.0000000	0.9976218	0.9982988	0.9999464
34	1.0000000	0.9976513	0.9984115	0.9999467
36	1.0000000	0.9976796	0.9985178	0.9999473
38	1.0000000	0.9977066	0.9986190	0.9999482
40	1.0000000	0.9977323	0.9987157	0.9999493
42	1.0000000	0.9977568	0.9988082	0.9999505
44	1.0000000	0.9977801	0.9988967	0.9999518
46	1.0000000	0.9978027	0.9989810	0.9999532
48	1.0000000	0.9978251	0.9990608	0.9999546
50	1.0001842	0.9978693	0.9991445	0.9999564

b. Ritz chamber (cont'd).

E/keV	k_{el}	k_{sc}	k_{fl}	k_{br}
52	1.0003224	0.9978882	0.9992122	0.9999576
54	1.0006910	0.9979065	0.9992744	0.9999588
56	1.0014978	0.9979257	0.9993314	0.9999598
58	1.0031364	0.9979465	0.9993837	0.9999606
60	1.0058425	0.9979695	0.9994315	0.9999614
62	1.0098340	0.9979947	0.9994753	0.9999620
64	1.0151261	0.9980219	0.9995153	0.9999625
66	1.0215274	0.9980505	0.9995521	0.9999628
68	1.0287504	0.9980795	0.9995860	0.9999630
70	1.0363823	0.9981077	0.9996173	0.9999631
72	1.0439026	0.9981340	0.9996463	0.9999631
74	1.0507998	0.9981575	0.9996734	0.9999629
76	1.0566715	0.9981775	0.9996987	0.9999626
78	1.0612800	0.9981938	0.9997224	0.9999623
80	1.0645351	0.9982071	0.9997446	0.9999619
82	1.0665470	0.9982190	0.9997654	0.9999615
84	1.0675789	0.9982316	0.9997847	0.9999611
86	1.0679969	0.9982478	0.9998027	0.9999608
88	1.0681445	0.9982705	0.9998192	0.9999605
90	1.0681937	0.9983018	0.9998343	0.9999605
92	1.0682183	0.9983417	0.9998478	0.9999605
94	1.0679231	0.9983867	0.9998596	0.9999608
96	1.0669646	0.9984280	0.9998697	0.9999611
98	1.0650010	0.9984495	0.9998777	0.9999615
100	1.0623803	0.9984232	0.9998837	0.9999617

c. Wyckoff-Attix chamber.

E/keV	k_{el}	k_{sc}	k_{fl}	k_{br}
3.206	1.0000000	0.9867256	0.9917779	0.9999849
5	1.0000000	0.9881827	0.9925562	0.9999820
10	1.0000000	0.9908061	0.9942537	0.9999727
20	1.0000000	0.9930521	0.9963685	0.9999534
30	1.0000000	0.9938706	0.9976026	0.9999399
40	1.0000000	0.9942853	0.9983913	0.9999347
50	1.0000000	0.9946267	0.9989182	0.9999357
60	1.0000000	0.9949880	0.9992746	0.9999392
70	1.0000000	0.9953677	0.9995147	0.9999430
80	1.0000000	0.9957372	0.9996746	0.9999458
90	1.0006910	0.9960689	0.9997799	0.9999472
100	1.0022591	0.9963514	0.9998488	0.9999472
110	1.0040839	0.9965870	0.9998940	0.9999459
120	1.0059584	0.9967829	0.9999237	0.9999436
130	1.0075347	0.9969484	0.9999435	0.9999406
140	1.0086489	0.9970918	0.9999568	0.9999371
150	1.0091832	0.9972195	0.9999659	0.9999334
160	1.0091135	0.9973355	0.9999722	0.9999299
170	1.0085792	0.9974420	0.9999767	0.9999268
180	1.0076971	0.9975401	0.9999799	0.9999243
190	1.0066767	0.9976312	0.9999823	0.9999224
200	1.0057036	0.9977169	0.9999841	0.9999211
210	1.0050785	0.9977993	0.9999856	0.9999205
220	1.0050323	0.9978800	0.9999868	0.9999203
230	1.0059815	0.9979596	0.9999878	0.9999204
240	1.0084399	0.9980361	0.9999887	0.9999207
250	1.0130479	0.9981051	0.9999895	0.9999208
260	1.0206339	0.9981597	0.9999902	0.9999205
270	1.0320720	0.9981945	0.9999908	0.9999196
280	1.0478521	0.9982107	0.9999915	0.9999177
290	1.0677756	0.9982288	0.9999922	0.9999148
300	1.0898080	0.9983084	0.9999931	0.9999110

Table 7. Estimated relative standard uncertainties due to smoothing of the Burns data adopted for the NIST FACs.

Correction Factor	Lamperti (10 kV – 50 kV)	Ritz (20 kV – 100 kV)	Wyckoff-Attix (50 kV – 300 kV)
k_{el}	0.02 %	0.06 %	0.04 %
k_{sc}	0.01 %	0.01 %	0.01 %
k_{fl}	0.01 %	0.01 %	0.01 %

Table 8. Estimated relative standard uncertainties inherent in the data adopted for calculating correction factors for the NIST FACs (combining results in Tables 2 and 7). The last row gives the estimate associated with the assumption of $k_{br}/(1-\overline{g}) = 1$.

Correction Factor	Lamperti (10 kV – 50 kV)	Ritz (20 kV – 100 kV)	Wyckoff-Attix (50 kV – 300 kV)
k_{el}	0.05 %	0.08 %	0.10 %
k_{sc}	0.03 %	0.03 %	0.03 %
k_{fl}	0.05 %	0.05 %	0.03 %
$k_{br}/(1\text{-}g)$	0.02 %	0.02 %	0.03 %

Table 9. Adopted correction factors for the NIST free-air chambers: traditional NIST beam qualities.

Beam Code	Lamperti			Ritz			Wyckoff-Attix		
	k_{el}	k_{sc}	k_{fl}	k_{el}	k_{sc}	k_{fl}	k_{el}	k_{sc}	k_{fl}
L10	1.00000	0.99790	0.99608						
L15	1.00000	0.99818	0.99662						
L20	1.00000	0.99835	0.99698	1.00000	0.99622	0.99549			
L30	1.00000	0.99860	0.99758	1.00000	0.99676	0.99635			
L40	1.00000	0.99874	0.99798	1.00000	0.99707	0.99694			
L50	1.00153	0.99884	0.99831	1.00000	0.99730	0.99740	1.00000	0.99308	0.99656
L80				1.00128	0.99756	0.99810	1.00000	0.99377	0.99750
L100				1.00531	0.99764	0.99838	1.00002	0.99407	0.99792
M20	1.00000	0.99851	0.99732	1.00000	0.99656	0.99598			
M30	1.00000	0.99874	0.99790	1.00000	0.99704	0.99681			
M40	1.00008	0.99883	0.99823	1.00000	0.99726	0.99729			
M50	1.00202	0.99888	0.99844	1.00000	0.99738	0.99760	1.00000	0.99328	0.99681
M60				1.00003	0.99752	0.99796	1.00000	0.99363	0.99730
M80				1.00487	0.99766	0.99844	1.00000	0.99409	0.99796
M100				1.00946	0.99780	0.99885	1.00003	0.99453	0.99855
M120							1.00034	0.99487	0.99889
M150							1.00133	0.99548	0.99937
M200							1.00414	0.99637	0.99975
M250							1.00628	0.99712	0.99993
M300							1.01227	0.99780	0.99998
H10	1.00000	0.99801	0.99629						
H15	1.00000	0.99847	0.99720						
H20	1.00000	0.99871	0.99777	1.00000	0.99696	0.99661			
H30	1.00000	0.99892	0.99849	1.00000	0.99744	0.99766			
H40	1.00089	0.99896	0.99889	1.00000	0.99762	0.99828			
H50	1.01727	0.99898	0.99909	1.00000	0.99770	0.99859	1.00000	0.99419	0.99820
H60				1.00035	0.99781	0.99899	1.00000	0.99452	0.99873
H100				1.05982	0.99824	0.99976	1.00045	0.99584	0.99970
H150							1.00575	0.99674	0.99991
H200							1.00815	0.99737	0.99997
H250							1.00633	0.99777	0.99998
H300							1.02266	0.99808	0.99999
S60				1.00005	0.99763	0.99835	1.00000	0.99399	0.99784
S75				1.00080	0.99755	0.99810	1.00000	0.99376	0.99750

Table 10. Adopted correction factors for the NIST free-air chambers: ISO beam qualities. The NS10, NS15, NS20, NS30, and NS40 beam qualities are assumed to be the same as that of the NIST H10, H15, H20, H30, and H40 beams, respectively.

Beam	Lamperti			Ritz			Wyckoff-Attix		
Code	k_{el}	k_{sc}	k_{fl}	k_{el}	k_{sc}	k_{fl}	k_{el}	k_{sc}	k_{fl}
HK10	1.00000	0.99790	0.99608						
HK20	1.00000	0.99841	0.99712	1.00000	0.99636	0.99568			
HK30	1.00000	0.99872	0.99786	1.00000	0.99700	0.99675			
HK60				1.00005	0.99760	0.99824	1.00000	0.99389	0.99769
HK100				1.01306	0.99789	0.99910	1.00004	0.99478	0.99888
HK200							1.00414	0.99637	0.99975
HK250							1.00541	0.99687	0.99987
HK280							1.00706	0.99724	0.99994
HK300							1.00854	0.99727	0.99993
WS60				1.00020	0.99776	0.99881	1.00000	0.99438	0.99850
WS80				1.00958	0.99792	0.99923	1.00000	0.99482	0.99904
WS110							1.00058	0.99570	0.99961
WS150							1.00400	0.99646	0.99984
WS200							1.00722	0.99708	0.99994
WS250							1.00716	0.99751	0.99997
WS300							1.01227	0.99780	0.99998
NS10	1.00000	0.99801	0.99629						
NS15	1.00000	0.99847	0.99720						
NS20	1.00000	0.99871	0.99777	1.00000	0.99696	0.99661			
NS25	1.00000	0.99884	0.99815	1.00000	0.99724	0.99715			
NS30	1.00000	0.99892	0.99849	1.00000	0.99744	0.99766			
NS40	1.00089	0.99896	0.99889	1.00000	0.99762	0.99828			
NS60				1.00035	0.99781	0.99899	1.00000	0.99452	0.99873
NS80				1.02295	0.99804	0.99951	1.00000	0.99517	0.99937
NS100				1.05982	0.99824	0.99976	1.00045	0.99584	0.99970
NS120							1.00260	0.99635	0.99984
NS150							1.00569	0.99674	0.99991
NS200							1.00814	0.99739	0.99997
NS250							1.00634	0.99779	0.99998
NS300							1.02266	0.99808	0.99999
LK10	1.00000	0.99810	0.99646						
LK20	1.00000	0.99878	0.99795	1.00000	0.99711	0.99685			
LK30	1.00000	0.99894	0.99860	1.00000	0.99750	0.99784			
LK35	1.00002	0.99895	0.99880	1.00000	0.99758	0.99813			
LK55				1.00012	0.99782	0.99901	1.00000	0.99452	0.99876
LK70				1.01031	0.99798	0.99942	1.00000	0.99501	0.99927
LK100				1.06452	0.99828	0.99980	1.00064	0.99596	0.99974
LK125							1.00405	0.99656	0.99989
LK170							1.00868	0.99719	0.99996
LK210							1.00703	0.99758	0.99998
LK240							1.00575	0.99782	0.99999

Table 11. Adopted correction factors for the NIST free-air chambers: BIPM/CCRI beam qualities. The BIPM10 beam quality is assumed to be the same as that of the ISO HK10 beam; the BIPM50b beam quality is assumed to be the same as that of the NIST M50 beam.

Beam Code	Lamperti			Ritz			Wyckoff-Attix		
	k_{el}	k_{sc}	k_{fl}	k_{el}	k_{sc}	k_{fl}	k_{el}	k_{sc}	k_{fl}
BIPM10	1.00000	0.99790	0.99608	1.00000	0.99523	0.99415			
BIPM25	1.00000	0.99858	0.99751	1.00000	0.99672	0.99625			
BIPM30	1.00000	0.99790	0.99608	1.00000	0.99523	0.99415			
BIPM50A	1.00530	0.99895	0.99883	1.00000	0.99749	0.99788			
BIPM50B	1.00202	0.99888	0.99844	1.00000	0.99738	0.99760	1.00000	0.99328	0.99681
BIPM100				1.00709	0.99773	0.99862	1.00002	0.99431	0.99824
BIPM135							1.00068	0.99520	0.99918
BIPM180							1.00281	0.99588	0.99951
BIPM250							1.00556	0.99691	0.99986

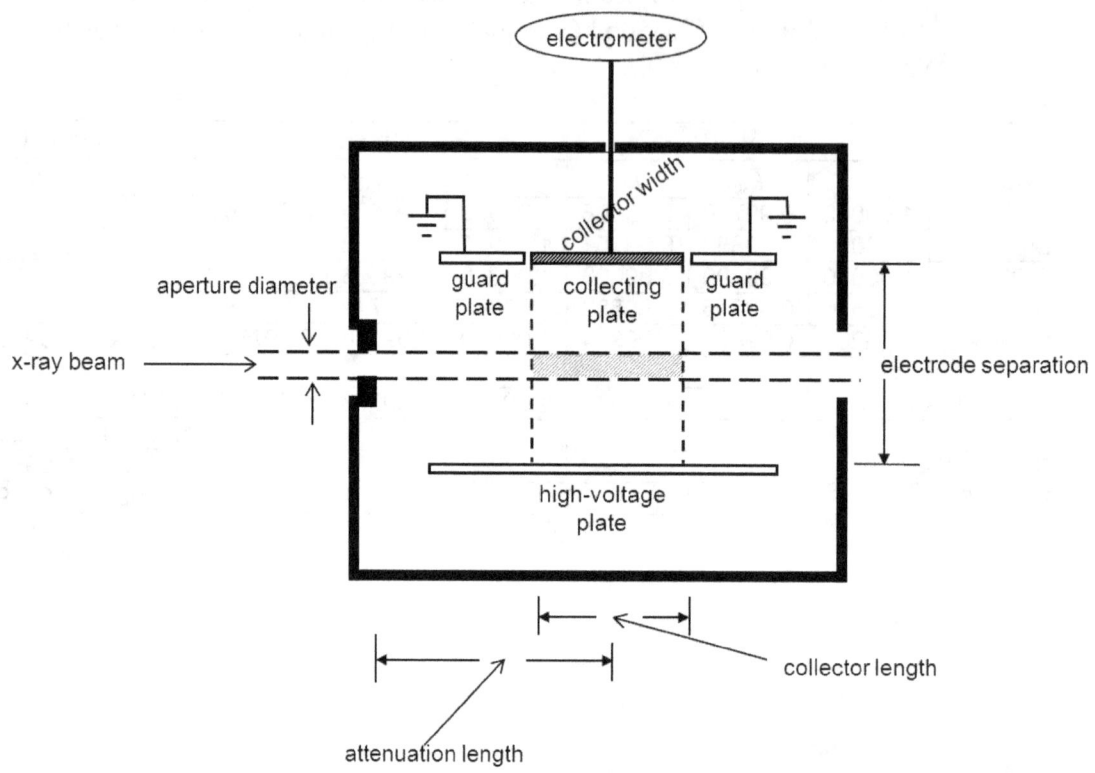

Figure 1. Schematic of a parallel-plate free-air ionization chamber used by the NIST for the realization of air kerma for W-anode x-ray beams. The high-voltage, guard, and collecting plates have a width that extends into the paper.

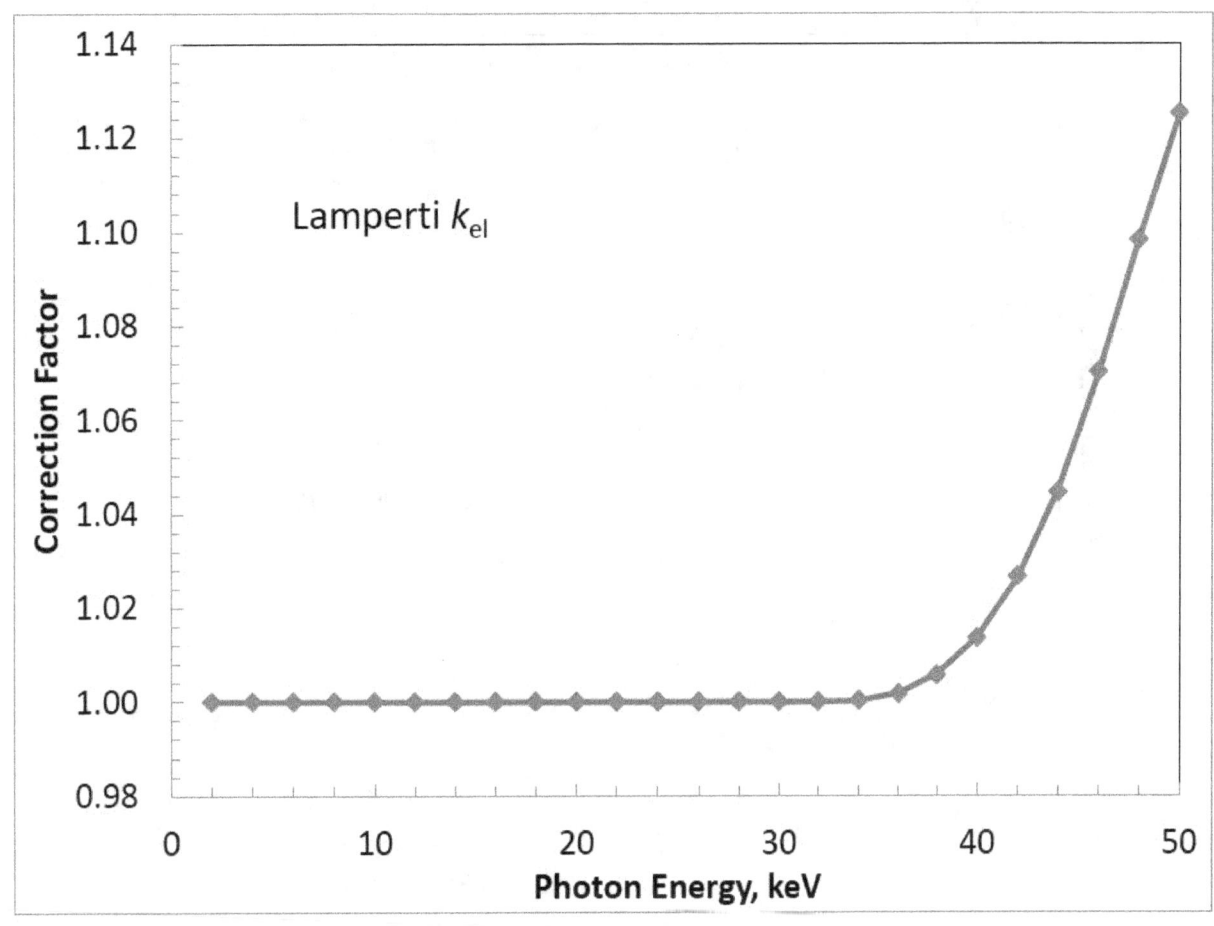

Figure 2. Comparison of original Burns data (points) for the Lamperti free-air chamber with the results from the smoothed underlying data (curve).

a. Correction factor k_{el}.

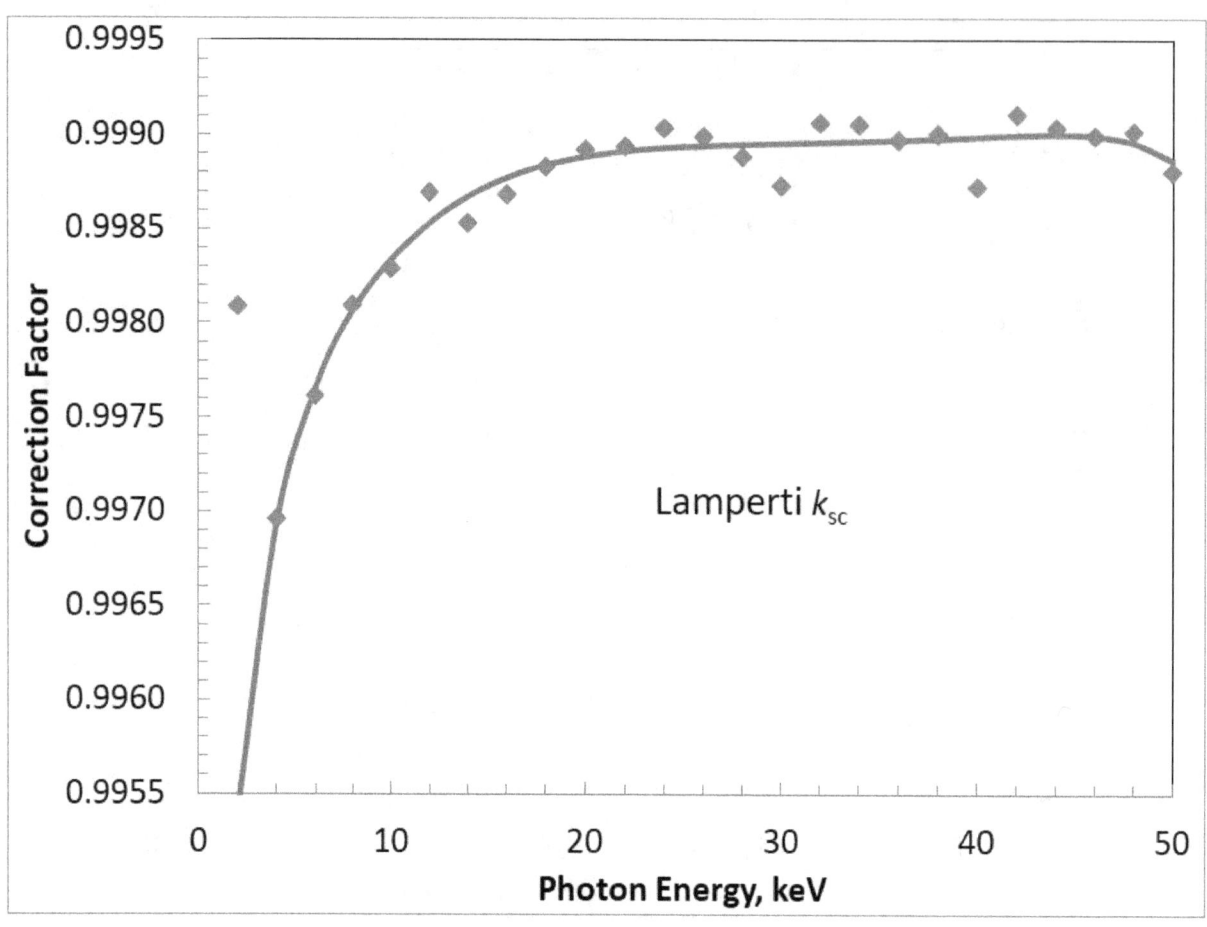

Figure 2. Comparison of original Burns data (points) for the Lamperti free-air chamber
with the results from the smoothed underlying data (curve).

b. Correction factor k_{sc}.

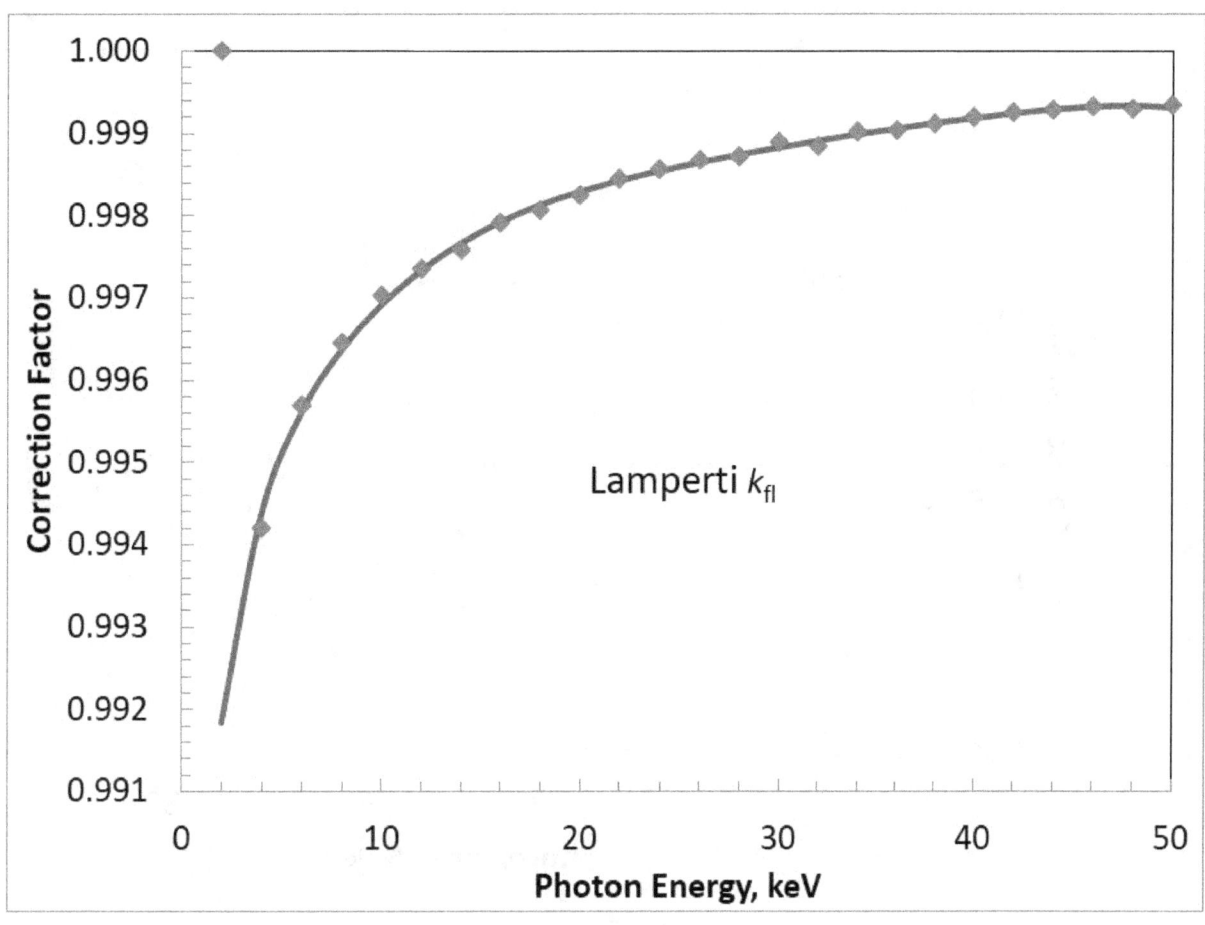

Figure 2. Comparison of original Burns data (points) for the Lamperti free-air chamber with the results from the smoothed underlying data (curve).

c. Correction factor k_{fl}.

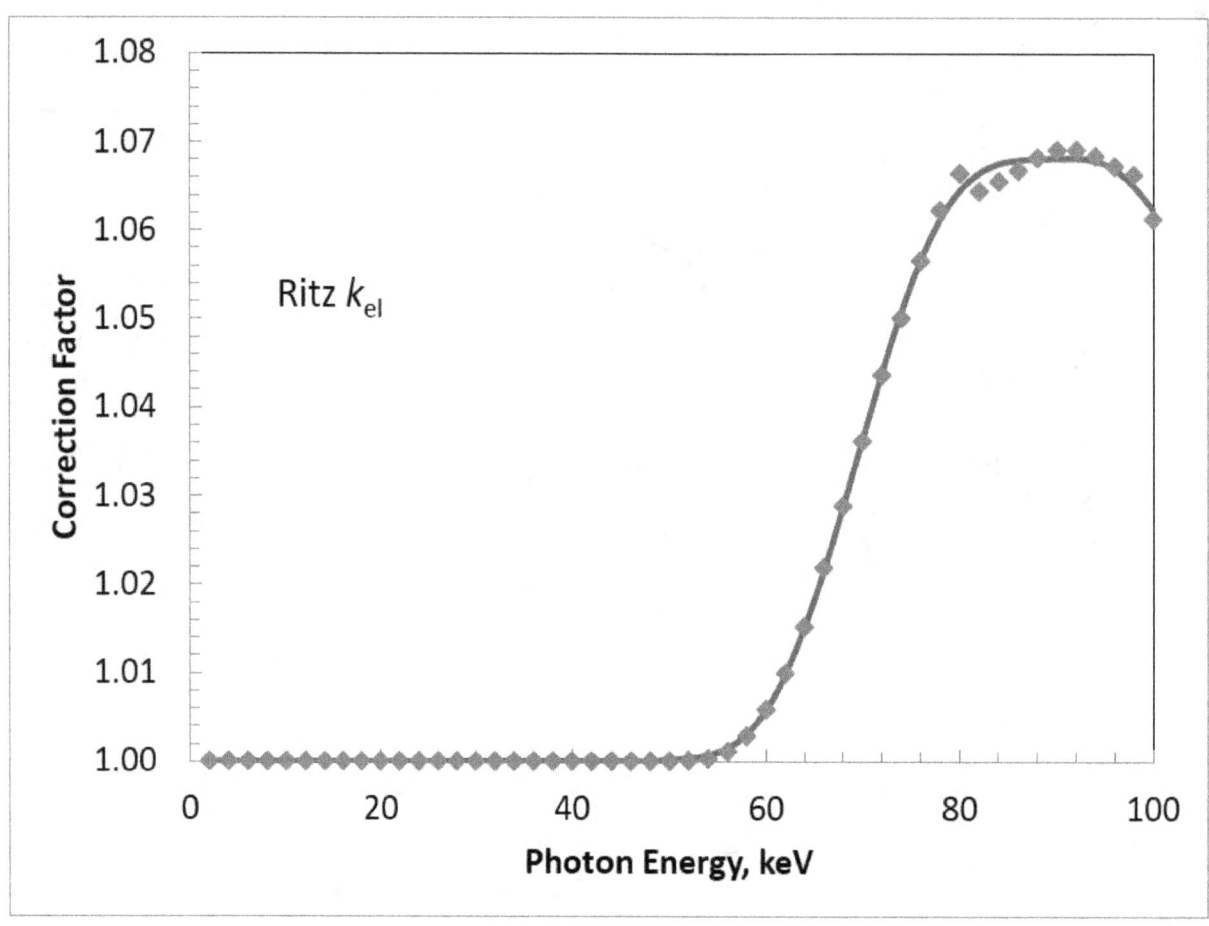

Figure 3. Comparison of original Burns data (points) for the Ritz free-air chamber with
the results from the smoothed underlying data (curve).

a. Correction factor k_{el}.

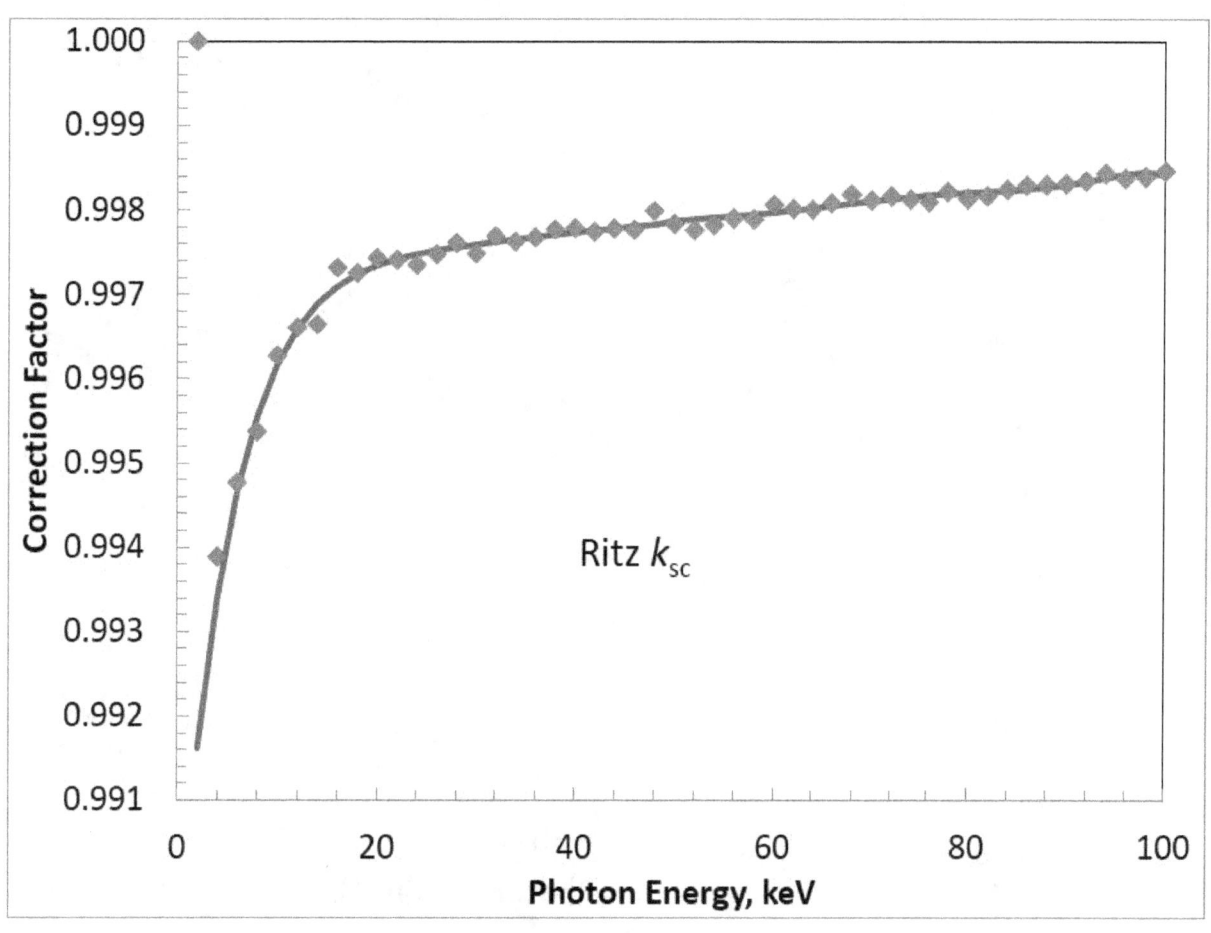

Figure 3. Comparison of original Burns data (points) for the Ritz free-air chamber with
the results from the smoothed underlying data (curve).

b. Correction factor k_{sc}.

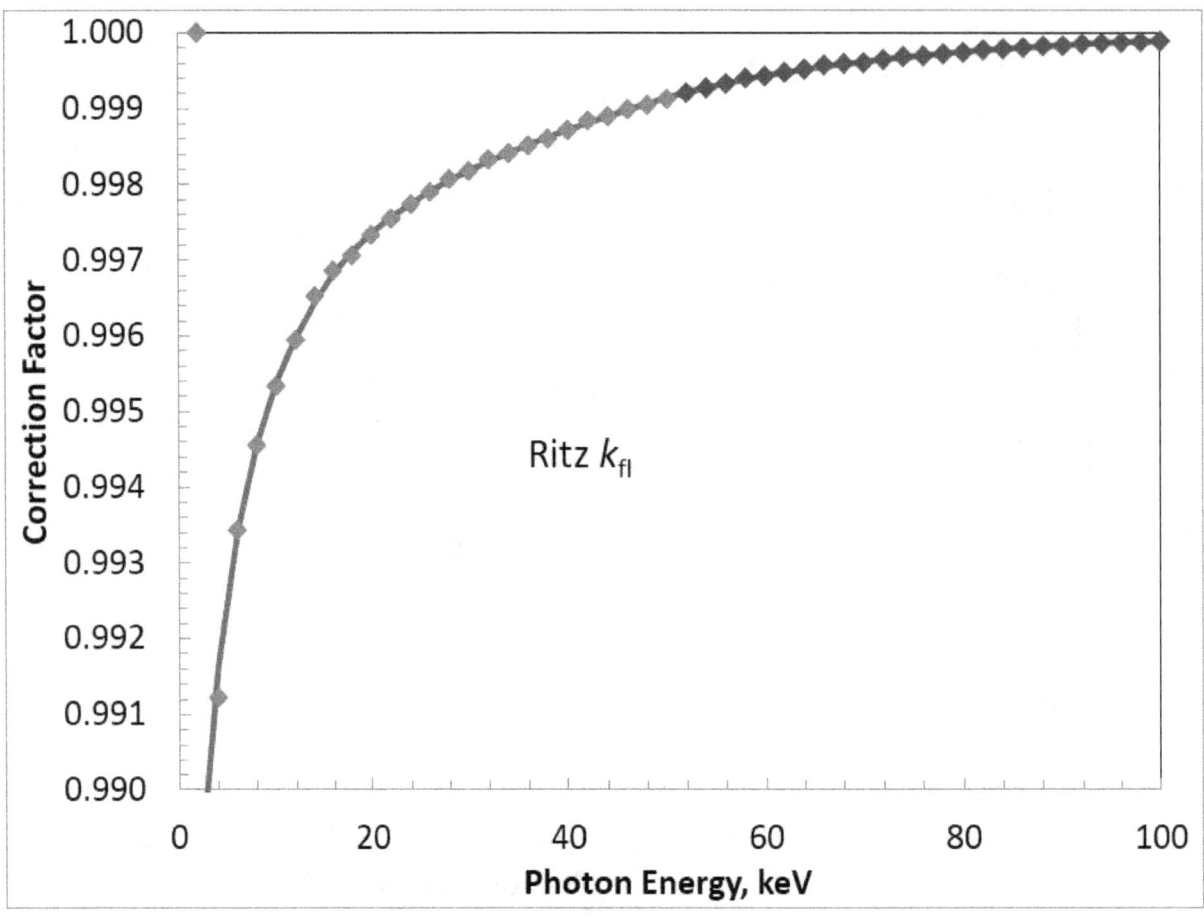

Figure 3. Comparison of original Burns data (points) for the Ritz free-air chamber with
the results from the smoothed underlying data (curve).

c. Correction factor k_{fl}.

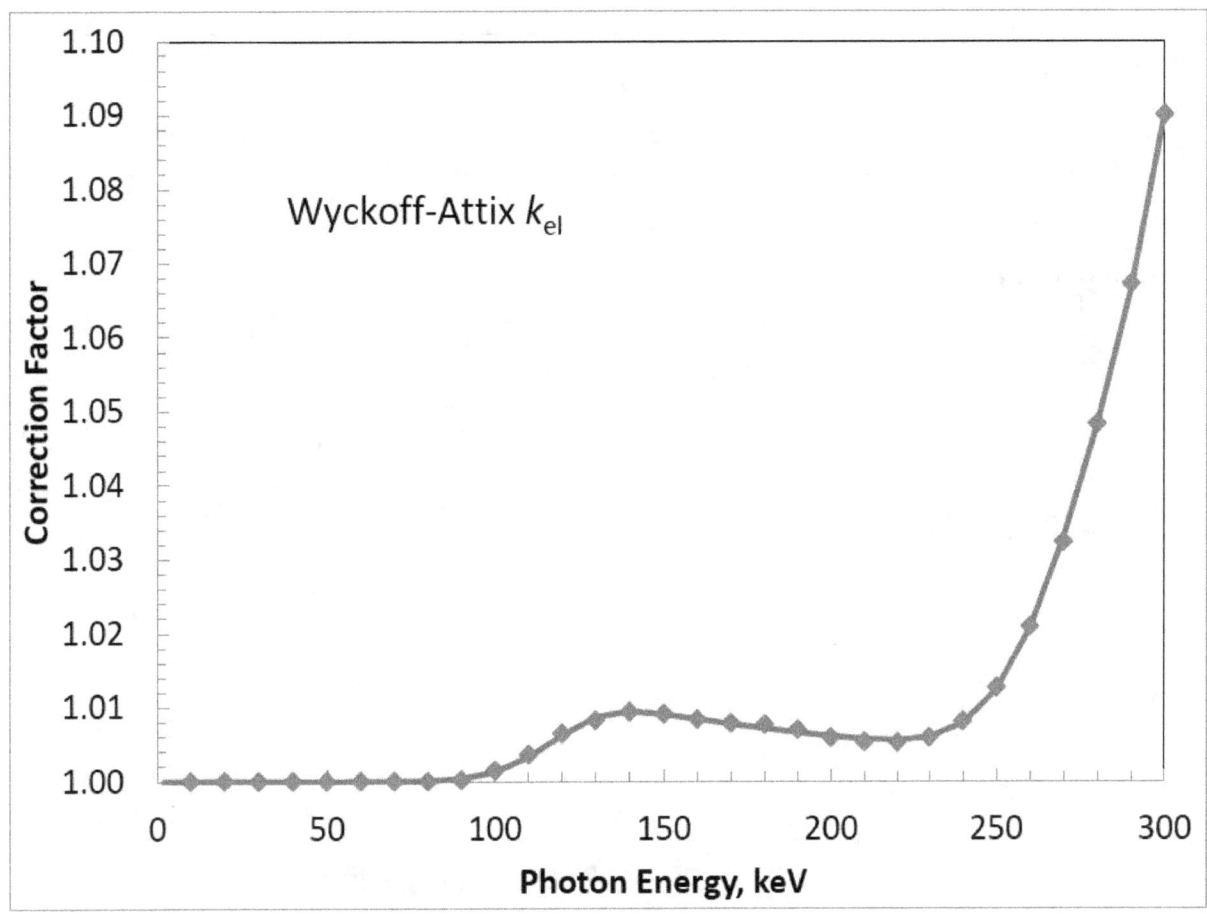

Figure 4. Comparison of original Burns data (points) for the Wyckoff-Attix free-air chamber with the results from the smoothed underlying data (curve).

a. Correction factor k_{el}.

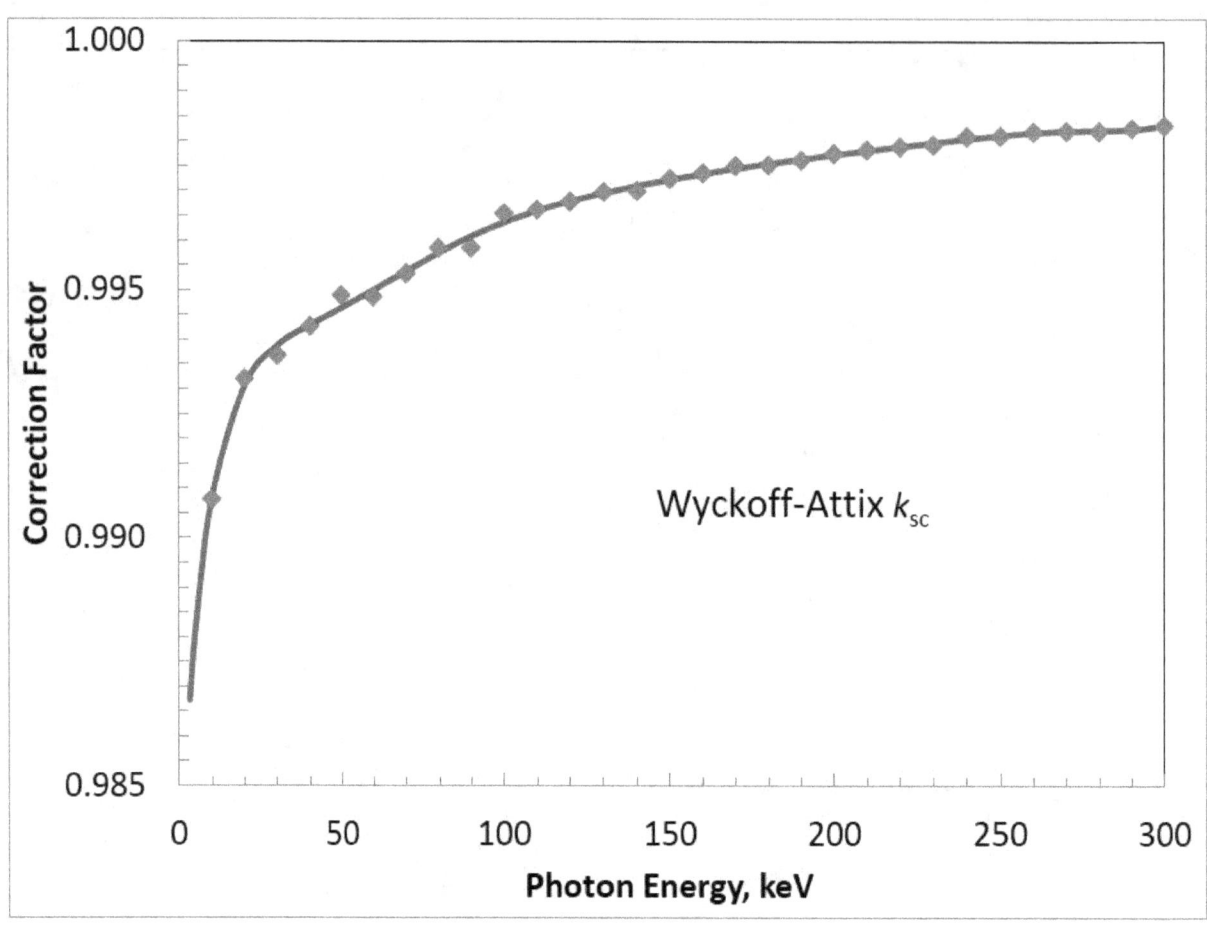

Figure 4. Comparison of original Burns data (points) for the Wyckoff-Attix free-air chamber with the results from the smoothed underlying data (curve).

b. Correction factor k_{sc}.

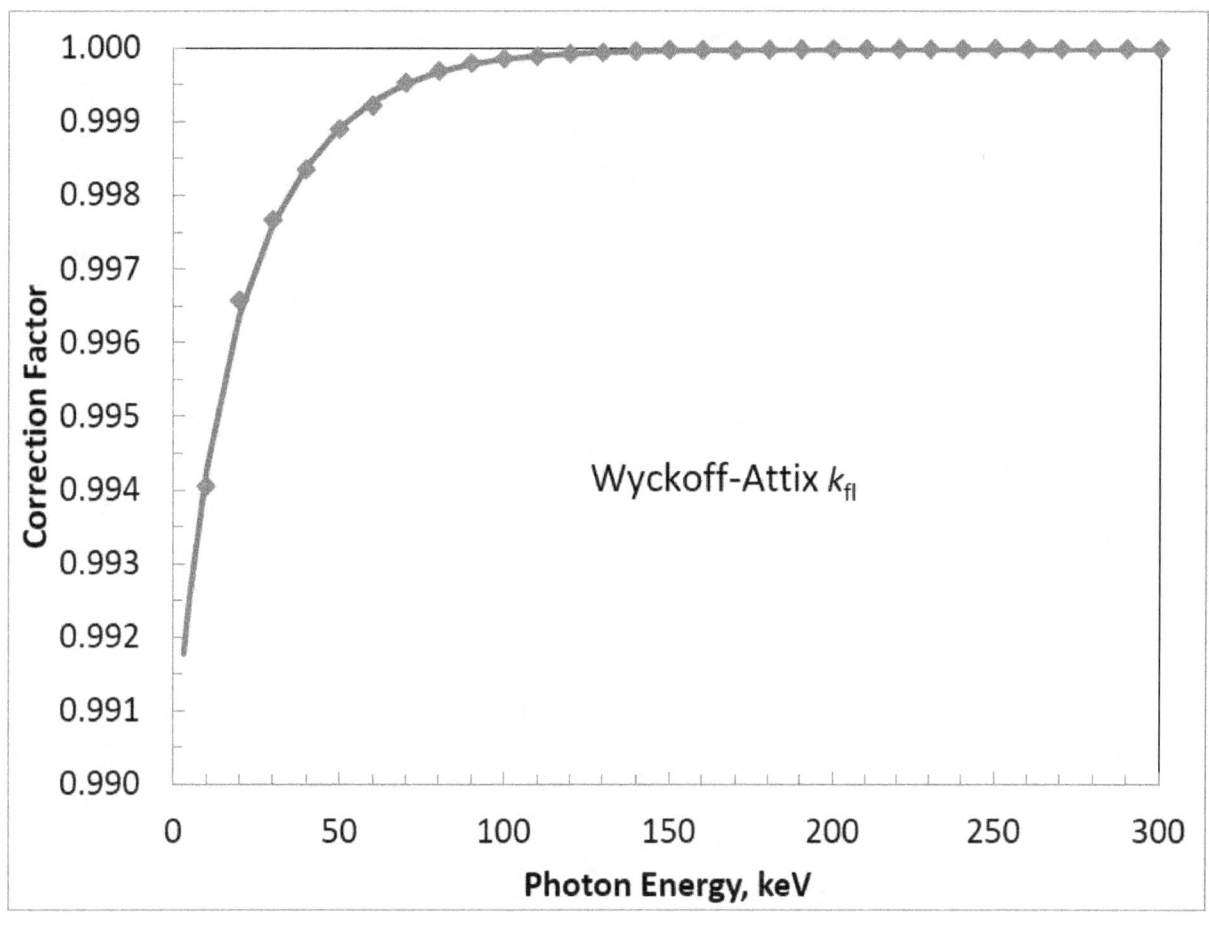

Figure 4. Comparison of original Burns data (points) for the Wyckoff-Attix free-air
chamber with the results from the smoothed underlying data (curve).

c. Correction factor k_{fl}.